Principles of Cellular, Molecular, and Developmental Neuroscience

Oswald Steward

Principles of Cellular, Molecular, and Developmental Neuroscience

With 128 Figures in 277 Parts

Springer-Verlag
New York Berlin Heidelberg
London Paris Tokyo

Oswald Steward, Ph.D.
Department of Neuroscience
University of Virginia
School of Medicine
Charlottesville, VA 22908, USA

Cover: A hippocampal neuron grown in culture. The neuron is immunostained with an antibody to the microtubule associated protein MAP2. Courtesy G.A. Banker.

Library of Congress Cataloging-in-Publication Data
Steward, Oswald.
    Principles of cellular, molecular, and developmental neuroscience/
Oswald Steward.
        p.   cm.
    Includes bibliographies and index.
    ISBN 0-387-96803-2
    1. Molecular neurobiology.   2. Nervous system—Cytology.
3. Developmental neurology.   I. Title.
    [DNLM: 1. Nervous System—cytology.   2. Nervous System—growth &
development.   3. Nervous System—physiology.   WL 102 S849p]
QP356.2.S74   1988
591.1′88—dc19
DNLM/DLC   88-20145

Typeset by David E. Seham Associates, Inc., Metuchen, New Jersey.
Printed and bound by Arcata Graphics/Halliday, West Hanover, Massachusetts.
Printed in the United States of America.

9 8 7 6 5 4 3 2 1

ISBN 0-387-96803-2 Springer-Verlag New York Berlin Heidelberg
ISBN 3-540-96803-2 Springer-Verlag Berlin Heidelberg New York

By rights, this book should be dedicated to my wife and children, since they suffered through its creation. However, since this book is my first, it is dedicated to my parents, Oswald Steward, Sr., and Ann Griffiths Steward, for their trust and support during much more difficult periods.

Preface

The field of cellular, molecular, and developmental neuroscience represents the interface between the three large, well established fields of neuroscience, cell biology, and molecular biology. In the last 10 to 15 years, this new field has emerged as one of the most rapidly growing and exciting subdisciplines of neuroscience. It is now becoming possible to understand many aspects of nervous system function at the molecular level, and there already are dramatic applications of this information to the treatment of nervous system injury, disease, and genetic disorders. Moreover, there is great optimism that new strategies will emerge soon as a result of the explosion of information.

This book was written to introduce students to the major issues, experimental strategies, and current knowledge base in cellular, molecular, and developmental neuroscience. The concept for the book arose from a section of an introductory neuroscience course given to first-year medical students at the University of Virginia School of Medicine. The text presumes a basic, but not detailed, understanding of nervous system organization and function, and a background in biology. It is intended as an appropriate introductory text for first-year medical students or graduate students in neuroscience, neurobiology, psychobiology, or related programs, and for advanced undergraduate students with appropriate background in biology and neuroscience. While some of the specific information presented undoubtedly will be outdated rapidly, the "gestalt" of this emerging field of inquiry as presented here should help the beginning student organize new information.

The book begins by considering the cellular and molecular biology of neurons and glia. Subsequent chapters review how neurons synthesize and process neurotransmitters, and the types of receptors and second-messenger systems that neurons possess. Three chapters are devoted to nervous system development and discuss nervous system histogenesis, growth of axons and dendrites, and synapse formation. The final chapter considers how the nervous system responds to injury. The selection of material presented reflects the biases and personal taste of the author. My apologies for the inevitable omissions and oversights.

Because the book is intended to provide an introduction to principles, citations within the text have been kept to a minimum. Nevertheless, to the extent possible, illustrations have been drawn from original experimental work. Suggested readings at the end of each chapter provide an additional guide to the primary literature. Particularly important "landmark" studies are included, as are recent reviews that will help students understand and appreciate the original experimental literature.

I am indebted to many colleagues who contributed to the development of this book with both specific advice and general encouragement. I especially thank my colleagues and collaborators Edwin Rubel and William Levy, who have had a tremendous influence on my thinking and writing. The chapter on histogenesis is a direct descendent of a lecture organized by Edwin Rubel for the "Fundamentals of Neuroscience" course at the University of Virginia. Special thanks also to Paula M. Falk for her help in preparing the illustrations and to Lauren Davis for helping with the final proofreading.

OSWALD STEWARD

Contents

CHAPTER 1

Cell Biology of Neurons

Part I
Neuronal Cell Structure and Its Relation to Function

Neurons from a Cellular Perspective

Certainly the most important function of the brain is the receipt, transmission, and processing of information. For this reason, much of our knowledge about neurons has to do with their signaling capabilities. Nevertheless, neurons are cells, and some of their most important functional properties arise from their cellular characteristics and from cell-cell interactions that are not directly related to signaling activities. For example, we tend to think of synapses as specializations for modulating activity of the receiving cell, that is, for the transmission of excitation or inhibition. Synaptic contacts play other important roles, however, in determining the structure, functional properties and very existence of the cells that they contact. These trophic or regulatory functions are of immense importance for understanding the consequences of injury or disease to the nervous system, since damage or disease at one site can result in a cascade of effects throughout the nervous system as a result of loss of intercellular regulation. This chapter considers neurons from a cellular perspective, focusing on those aspects of cellular existence that are unique in neurons or that figure importantly in neuronal function.

In Part I, we will consider the structural features of neurons. Special attention is given to those structural features that are directly related to physiological function. Part II will consider the molecular anatomy of neurons, intracellular transport processes, and the regulation of neuronal gene expression.

The Cytology of the Nerve Cell

A distinguishing characteristic of neurons is their complex and highly differentiated form. In general, neurons are categorized according to their processes (either axonal or dendritic). The most simple classification is based on the number of processes that a cell emits. Thus, cells that give rise to a single process are termed *unipolar;* cells that give rise to two processes are *bipolar;* and cells that give rise to multiple processes are *multipolar.* Unipolar cells are rare except during early development, when migrating cells often give rise to a single leading process. In mature animals, spinal ganglion cells are unipolar, having a single process that later bifurcates to give rise to a centrally projecting axon and a peripherally

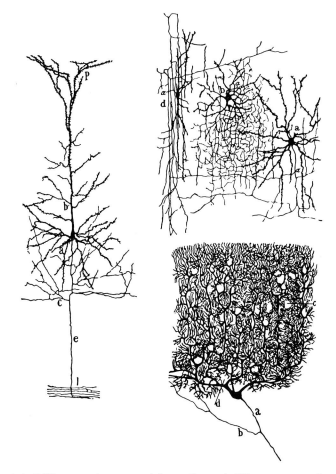

FIGURE 1.1. Differences in neuronal form. Several different types of neurons are illustrated. On the left is a pyramidal cell of the cerebral cortex; above right are several examples of short axon (Golgi type II cells) in the cerebral cortex; below right is a cerebellar Purkinje cell with its extensive dendritic tree that arborizes in a planar projection. [From Ramon y Cajal S *Histologie du Système Nerveux de l'Homme et des Vertébrés* (1911). Consejo Superior de Investigaciones Cientificas, Madrid, 1955.]

projecting sensory nerve. Nevertheless, these cells are bipolar during early development, after which their two principal processes fuse. For this reason, spinal ganglion cells are termed *pseudounipolar*.

The fact that most neurons are bipolar or multipolar is a reflection of the *principle of dynamic polarization*. Most neurons have two types of processes—*axons,* which are specialized for the transmission of information over distances, and *dendrites,* which are specialized for the receipt of information. As will be seen, these two types of processes have quite different properties.

Most neurons in the central nervous system (CNS) are multipolar and exhibit a complex form that is highly characteristic for the cell type (Fig. 1.1). Differences in form thus provide a convenient means for classifying cell types, and at least some aspects of neuronal form are directly related to functional properties. As will be seen, cell types are distinguished on the basis of characteristics of either dendrites or axonal arborizations.

Despite the wide variance in their physical appearance, almost all neurons have an underlying similarity in their cellular organization. In the following sections, the commonalities among nerve cells are considered.

The Cell Body or Soma

The neuronal cell body or *soma* contains the nucleus, which is surrounded by cytoplasm containing most of the organelles that are present in other types of cells (Fig. 1.2). The cell body also serves as the point of origin for the axons and dendrites. Thus, the cell body serves as the cytocenter from which the complicated three-dimensional architecture of the neuronal cytoplasm and cytoskeleton is constructed and maintained.

The Nucleus

A substantial proportion of the genome is continually transcribed in neurons. In fact, based on hybridization studies, it is estimated that about one third of the genome is actively transcribed in brain. Given the immense degree of structural specialization of neurons, this is not surprising. Because of the high level of transcriptional activity, neurons are *euchromatic,* that is, the nuclear chromatin is dispersed. This is in contrast to glial cells, which usually have clumps of chromatin on the internal face of the nuclear membrane. The neuronal nucleus contains one or two prominent nucleoli, which often have attached electron dense bodies. One of the dense bodies often associated with the nucleoli is the *Barr body,* which is the condensed female sex chromatin.

The Cytoplasm of the Neuron

For all neurons, polyribosomes and rough endoplasmic reticulum (RER) are localized primarily in the cell body. Axons have essentially no ability to synthesize protein, since they do not contain ribosomes or significant

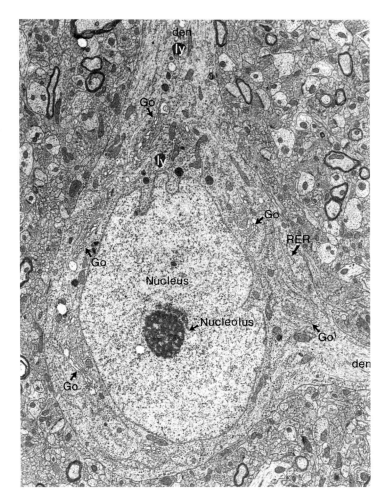

FIGURE 1.2. Electron micrograph of the cell body of a cortical pyramidal cell (rat) showing the typical constituents of the neuronal cytoplasm. Dendrites (den) extend from the cell body. ly = lysosomes, Go = Golgi apparatus; RER = rough endoplasmic reticulum.

amounts of RNA, except within the initial segment (see below). Thus, axons depend entirely on proteins that are produced in the cell body. Dendrites do contain both RNA and ribosomes (see below), and there is reason to believe that this protein synthesis machinery plays an important role in dendritic function. Nevertheless, most of the protein constituents of both axons and dendrites must be synthesized in the cell body, and thus axons and dendrites depend on the cell body for metabolic support. For this reason, the cell body was termed the *trophic center of the neuron* by Santiago Ramon y Cajal.

The machinery for protein synthesis within neuronal cell bodies includes stacks of rough endoplasmic reticulum, large numbers of polyribosomes (collections of individual ribosomes attached to an individual molecule of messenger ribonucleic acid, mRNA) and free monosomes. Because of the large concentrations of ribosomes, neuronal cell bodies stain heavily with basic dyes such as thionin and cresyl violet. In many of the larger neurons, thionin and cresyl violet stains reveal clumps of granules that are called *Nissl bodies*, after Franz Nissl (1860–1919), who initially described them. These Nissl bodies represent the endoplasmic reticulum visible at the electron microscopic level. Histological stains using basic dyes that reveal the Nissl bodies are termed Nissl stains.

The distribution of Nissl substance reveals the protein synthetic activity of the neuron. For example, when neurons are *axotomized*, they often exhibit *chromatolysis*, which is characterized by a dispersal of Nissl substance throughout the cytoplasm. This chromatolytic response provided one of the earliest ways to trace pathways in the CNS, since cells whose axons are damaged by a lesion exhibit chromatolysis (the retrograde degeneration method of Gudden). Similar changes in Nissl staining are observed in neurons undergoing transneuronal degeneration after deafferentation. For example, within hours after the cessation of activity over eighth nerve axons in the chick, neurons in the auditory brainstem undergo a transneuronal degeneration. During the early phase of degeneration, the cells cease to stain with Nissl stains, appearing as "ghost cells." Simultaneous with the decrease in Nissl staining, the cells cease protein synthesis (Fig. 1.3). Thus, during both retrograde and transneuronal degeneration, changes in the amount or distribution of Nissl substance indicate substantial changes in neuronal protein synthesis.

Neuronal cell bodies also contain a prominent Golgi apparatus, as do other secretory cells. Often, the Golgi apparatus can be found at the base of dendrites. As in other cell types, the Golgi apparatus of neurons is engaged in the terminal glycosylation of proteins that are synthesized on the rough endoplasmic reticulum. Presumably, many or all of the membrane proteins and the proteins destined for release from the cell undergo terminal glycosylation in the Golgi apparatus. This would include the proteins of the neuronal plasma membrane (including the specialized proteins, such as the ion channels, receptors, ionophores, etc., involved in the physiological function of neurons) as well as the peptides released by some neurons. Since the Golgi apparatus is found only in the cell body, some means is required to deliver these proteins to their eventual site of utilization. This requires a highly specialized intracellular transport system (see below).

It is thought that the proteins of the axonal cytoskeleton are synthesized upon free polyribosomes that accumulate around the axon's point of origin. Membrane proteins and proteins for release are, in general, synthesized on rough endoplasmic reticulum and processed through the Golgi apparatus

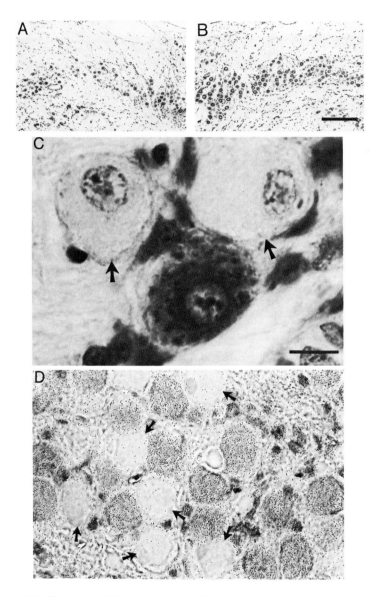

FIGURE 1.3. Changes in Nissl substance indicate changes in protein synthetic activity in neurons. Deafferentation of neurons in the brainstem auditory nuclei leads to the degeneration of the deafferented neurons (compare *A* deafferented with *B* control.) Early in the course of degeneration, there is a dramatic decrease in Nissl staining of the cells destined to die (*arrows* in *C*). Protein synthetic activity of the neurons can be evaluated by injecting radiolabeled precursors (3H-leucine) and preparing brain sections for autoradiography. In such preparations the neurons that exhibit decreases in Nissl staining are unlabeled and are thus not actively synthesizing protein (*arrows* in *D*), whereas neurons that exhibit normal staining

(for terminal glycosylation). After processing, they somehow become associated with the transport machinery of axons or dendrites, perhaps in the form of vesicles. How material is routed differentially into axons and dendrites is currently unknown.

Neuronal cell bodies also contain the other organelles that are found in other cell types such as mitochondria, lysozomes, and vesicular structures, including coated vesicles. Although special roles for these organelles in neurons have been suggested, it is likely that their primary role is identical in all cell types.

Axons

Axons are specialized for transmitting activity in a nondecremental fashion over considerable distances. In fact, single axons of some vertebrates can reach meters in length (consider, for example, the axons of the corticospinal tract of a giraffe). Using the Golgi stain he developed in the late 19th century, Camillio Golgi recognized two general types of cells based upon their axonal projections. Neurons that project out of the region containing the cell body (projection neurons) are known as *Golgi type I;* neurons that project locally in the region containing the parent cell body are known as *Golgi type II;* these are called *interneurons.*

Neurons typically give rise to only one axon. The axon may branch extensively throughout its length, but, generally, all branches originate from a single stem process. Branching occurs through the formation of *collaterals,* where the daughter branch is often approximately the same caliber as the parent. Some cells give rise to collateral branches that project back to the region of the cell body (recurrent collaterals). These recurrent collaterals often terminate on interneurons (usually inhibitory), providing a substrate for recurrent inhibition. Occasionally, recurrent collaterals may terminate upon the same cell that gives rise to the axon, forming *autapses.* With the exception of the *initial segment* and the *terminal arborization* (considered below), axons have fairly uniform structural characteristics throughout their length. Furthermore, except in terms of their course and site of termination, the axons of different types of neurons are quite similar in general appearance.

Some of the characteristic features of axons can best be appreciated

◁———————————————————————————————————————

are heavily labeled. [*A-C* are from Born DE, Rubel EW (1985) Afferent influences on brain stem auditory nuclei of the chicken: neuron number and size following cochlea removal. *J Comp Neurol* 231: 435–445. *D* is from Steward O, Rubel EW (1985) Afferent influences on development of the brain stem auditory nuclei of the chicken: cessation of amino acid incorporation as an antecedent of age-dependent transneuronal degeneration. *J Comp Neurol* 231: 385–395.]

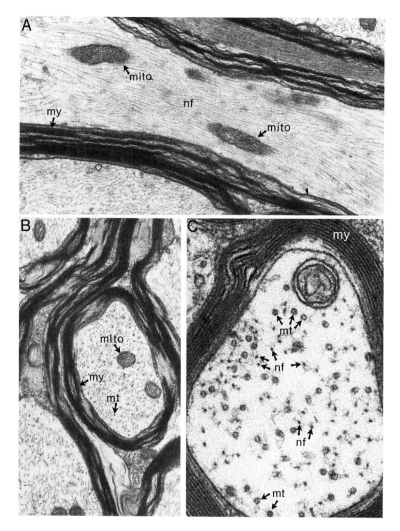

FIGURE 1.4. Electron micrographs of axons in the CNS. *A* Axon from brainstem of chicken sectioned longitudinally. *B* Axon from brainstem of chicken in cross-section. Note that in these axons there are very few microtubules (mt), but many neurofilaments (nf). *C* Axon from optic nerve of the rat. Microtubules and neurofilaments can be seen in cross-section. mito = mitochondria; my = myelin. (*A* and *B* courtesy of O. Steward and E.W. Rubel; *C* courtesy of P. Trimmer.)

by contrasting them with dendrites. First, unlike dendrites, which decrease in diameter with distance from the cell body, individual axons have a more or less uniform diameter throughout their length, at least until the axon breaks up into its terminal arborization. Second, whereas dendritic branches can occur at a variety of angles, axons tend to branch at right

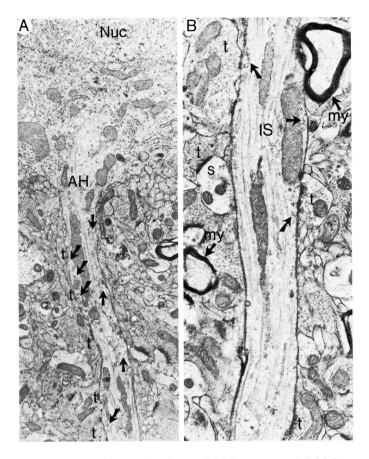

FIGURE 1.5. Electron micrographs of axon initial segments. *A* Initial segment of an axon of a cortical pyramidal cell in the rat. AH = axon hillock, Nuc = nucleus, t = terminals that contact the initial segment. *Arrows* indicate polyribosomes that are present in the initial segment. *B* Initial segment (IS) at higher magnification. my = myelin sheaths of nearby axons; s = spine synapse. Note the terminals (t) forming symmetric contacts with the initial segment. Many or all of these synapses have been shown to contain glutamic acid decarboxylase (GAD), the synthetic enzyme for GABA, and are thus thought to be inhibitory. [From Steward O, Ribak CE (1986) Polyribosomes associated with synaptic specializations on axon initial segments: localization of protein synthetic machinery at inhibitory synapses. *J Neurosci* 6: 3079–3085. Copyright, Society for Neuroscience.]

angles. Finally, the branching pattern and distribution of axonal processes depend on interactions between the axon and its surroundings. This is in contrast to dendrites, where the pattern and distribution of the dendritic arbor depend on processes that are intrinsic to the neuron (see Chapter 6).

Except at the initial segment and terminal arborization, axons exhibit few structural specializations throughout their length. In large-caliber axons, the most prominent organelles are longitudinally oriented cytoskeletal elements such as neurofilaments and microtubules. In most axons, neurofilaments predominate (for an extreme example, see Fig. 1.4). Mitochondria and membrane vesicles of various sizes are also common. Ribosomes are rarely present except in the initial segment of some axons (see below).

The *initial segment* begins at the axon's point of origin on the neuronal cell body *(the axon hillock)* and extends to the beginning of the myelin sheath (Fig. 1.5). A characteristic feature of the initial segment is a submembranous electron dense undercoating (Fig. 1.5B). A similar undercoating is present under nodes of Ranvier. Both of these sites contain high concentrations of sodium channels, and it has been suggested that the submembranous undercoating might be associated with these channels. The initial segments of many neurons of the CNS are also the site of synaptic connections (Fig. 1.5). Immunocytochemical studies have revealed that many of these synapses contain *glutamic acid decarboxylase,* the synthetic enzyme of the inhibitory neurotransmitter γ-aminobutyric acid (GABA). Thus, most of the connections on axon initial segments are thought to be inhibitory. Connections at this site should be highly effective in preventing transmission along the axon. Interestingly, there is a loss of these inhibitory connections in epileptic foci that are experimentally induced, suggesting that their loss may lead to uncontrolled epileptiform activity.

Terminal Specializations of Axons

Upon reaching the target area, axons usually exhibit some form of specialized branching pattern; this is termed the *preterminal arborization.* Many of these arborizations are highly characteristic. For example, the "basket" plexus formed by inhibitory interneurons on many types of cortical cells is highly stereotyped and, in fact, forms the basis for classifying the cells producing the basket plexus (Fig. 1.6).

The morphology of *presynaptic terminals* is also highly variable between cell types. In general, presynaptic terminals contain either vesicles (if the synapse is chemical) or membrane specializations appropriate for electrical synapses (the latter being rare in vertebrates). The nature of the presynaptic specialization is highly variable, however. Some take the form of *synaptic terminals* or *terminal boutons.* This term implies that the presynaptic specialization is at the terminus of the axon. A common type of synapse in cortical areas including the cerebellar cortex is the *en passant* variety, where presynaptic specializations are distributed along a length of the presynaptic axon (Fig. 1.7). Presynaptic terminals also differ greatly in size and complexity, ranging from simple processes that contain a few vesicles to immense and highly differentiated structures. Examples of the

FIGURE 1.6. The axonal plexus of a basket cell in the cerebellum. These neurons form characteristic basket endings around the cell bodies of Purkinje cells. [From Ramon y Cajal S *Histologie du Système Nerveux de l'Homme et des Vertébrés* (1911), Consejo Superior de Investigaciones Cientificas, Madrid, 1955.]

latter include cerebellar mossy fibers, *calyces of Held* in the anteroventral cochlear nucleus (AVCN, see Fig. 1.7C), and the large synaptic contacts in ciliary ganglia. Some neurons produce different types of contacts at different sites of termination. For example, eighth nerve afferents form calyces of Held on the cell bodies of neurons in the AVCN, but at sites of termination on dendrites they form simple synaptic knobs. This suggests that the form of the presynaptic contact must be determined by interactions between the presynaptic cell and its target.

Dendrites

Several features of dendrites distinguish them from axons, and it is useful to describe dendrites in terms of these differences. Dendrites taper throughout their length, and dendritic diameter is reduced in a predictable fashion at branch points. In contast, axonal diameter is roughly constant throughout its length. Because of the passive electrical properties of dendrites, diameter interacts with length in determining the integrative properties (see below). The principal structural features of dendrites are presumably related to the fact that dendrites are specialized as the principal receptive surface of the neuron.

As noted above, neurons typically have only one axon, but may have many dendrites. The number and, to some extent, the branching pattern of dendrites is characteristic of the neuron type. Indeed, dendritic form can often be used to identify neurons from different brain regions. Because most dendrites conduct electrotonically rather than via action potentials, dendritic form has important consequences for function (see below). De-

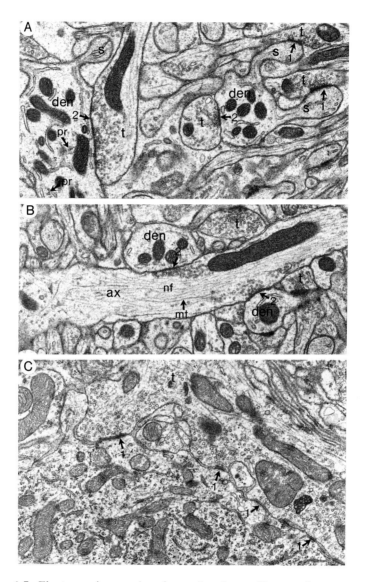

FIGURE 1.7. Electron micrographs of type I and type II synaptic contacts (rat cerebellum) *A* Type 1 (asymmetric) contacts are characterized by a dense post-synaptic membrane specialization (PSD); type 2 (symmetric contacts) have thin PSDs. Synapses on spines are almost always type 1, whereas synapses on dendritic shafts can be either type 1 or 2. den = dendrite; s = spine; t = presynaptic terminal; pr = polyribosomes. *B* Axon making *en passant* type 2 synapses with dendrites in the cerebellum. nf = neurofilaments; mt = microtubules. *C* Large synaptic ending forming a calyx of Held in the avian homologue of the anterior ventral cochlear nucleus. This synapse forms numerous type 1 synaptic junctions. (*C* is courtesy of O. Steward and E.W. Rubel.)

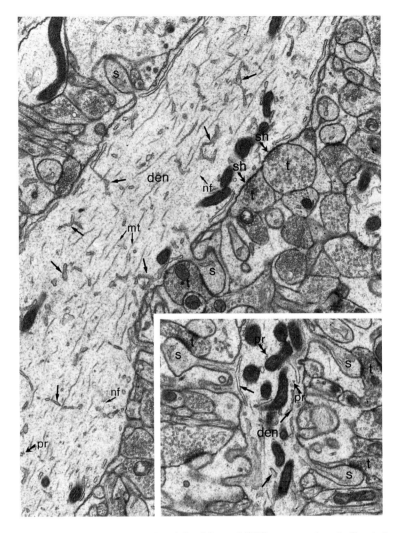

FIGURE 1.8. Electron micrographs of dendrites of CNS neurons (cerebellum). Large proximal dendrite of a Purkinje cell fills most of the field. Inset shows a thinner distal dendritic segment. den = dendrite; s = spine; sh = shaft synapse; t = presynaptic terminals; mt = microtubules; pr = polyribosomes. Note extensive cisternae of endoplasmic reticulum in dendrites (unlabeled *arrows*).

spite the wide variance in their physical appearance, dendrites do have an underlying similarity in their cellular organization.

Dendrites contain essentially all of the organelles found in axons and also ribosomes. Typically, the mitochondria in dendrites are very elongated, and the cytoskeletal elements, including neurofilaments and mi-

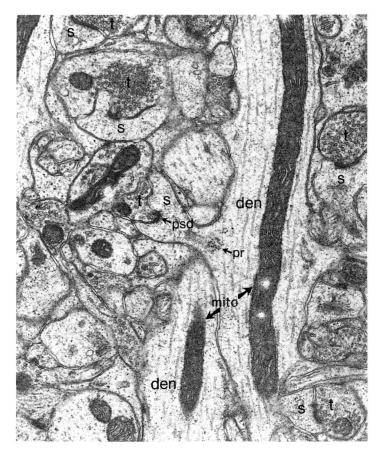

FIGURE 1.9. Electron micrographs of dendrites of CNS neurons (hippocampus). den = dendrite; s = spine; t = presynaptic terminal; psd = postsynaptic density; pr = polyribosomes; mito = mitochondria. [From Steward O (1983) Alterations in polyribosomes associated with dendritic spines during the reinnervation of the dentate gyrus of the adult rat. *J Neurosci* 3: 177–188. Copyright, Society for Neuroscience.]

crotubules, are oriented parallel to the long axis of the dendrite. In contrast to axons, microtubules are usually more numerous in dendrites than neurofilaments. Saccules of smooth endoplasmic reticulum are also present, more so in some neurons than others. For example, Purkinje cells of the cerebellum have extensive submembranous cisternae (Fig. 1.8), whereas the cisterns in cells of the cortex are less extensive (Fig. 1.9).

Synaptic Sites on Dendrites

Dendrites possess an enormous number of synaptic specializations along their length. In early electron microscopic studies, E.G. Gray recognized two general types of synapses. Asymmetric or Gray type I synapses have a dense postsynaptic membrane specialization (see Figs. 1.7 and 1.8). Symmetric or Gray type II synapses have thin postsynaptic densities (PSDs). The PSD is thought to represent the active zone of the synapse. It presumably contains neurotransmitter receptors, ionophores (channels specific to particular ions), recognition molecules (to determine the specificity of the contact), adhesion molecules (to promote attachment of the presynaptic process), and a variety of other molecules involved in synaptic function, including protein kinases, G proteins, calmodulin, and structural proteins to link the functional molecules to the site. In general, type I (asymmetric) synapses are thought to be excitatory, whereas type II synapses are thought to be inhibitory. This conclusion is based partly on studies that show that synapses containing the inhibitory neurotransmitter GABA form type II contacts (see Chapter 3).

Dendritic Spines

Many dendrites of CNS neurons also have microspecializations of form (dendritic spines) at each site of synpatic contact. Most neurons of the forebrain receive at least some synaptic contacts on spines, and many neurons, particularly in the cortex and cerebellum, receive virtually all

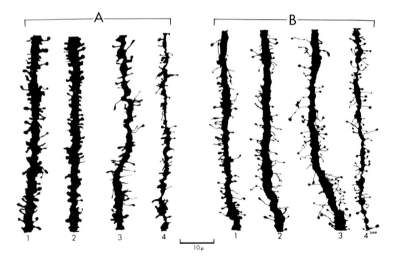

FIGURE 1.10. Spine abnormalities in retarded children. *A* Spine types from a normal 6-month-old infant. Dendrites 1 and 2 illustrate proximal dendritic segments with a predominance of stubby and mushroom-shaped spines, 3 illustrates spines from distal segments, and 4 illustrates a basal dendrite. *B* Similar sample of dendrites from a 10-month-old retarded child. Note overall spine density is less in the samples from the retarded child; also many abnormally long thin spines with large heads are present. [From Purpura DP (1979) Pathobiology of cortical neurons in metabolic and unclassified amentias, in Katzman R (ed): *Congenital and Acquired Cognitive Disorders.* New York, Raven Press, pp 43–68.]

of their *excitatory* synaptic connections on spines. Spines usually have type I (asymmetric) synaptic contacts, with prominent PSDs. Contacts on dendritic shafts may be either type I or II, however (see Figs. 1.7–1.9).

The size and length of spines are thought to vary as functions of the presence and activity of the presynaptic terminal. Quantitative studies of spines using the Golgi method, which stains individual neurons in their entirely, have revealed that deafferentation leads to a loss of spines, and the spines are replaced if the neurons are reinnervated (see Chapter 8). Conditions that lead to long-term decreases in activity are also correlated with spine loss. For example, visual deprivation leads to profound decreases in dendritic spines on pyramidal cells in the visual cortex. The number of spines also decrease in aging. Some types of mental retardation (Down's syndrome, for example) are characterized by profound spine loss in the cerebral cortex, as well as by abnormalities in the form of those spines that survive (Fig. 1.10). These abnormalities in spines profoundly alter the functional properties of spines because of the electrotonic properties that derive from the spine shape (see below).

An interesting feature of the ribosomes in dendrites is that they are usually selectively positioned beneath synaptic sites. In spine-bearing neurons the polyribosomes are found at the base of the spines (Fig. 1.9). The polyribosomes are particularly prominent during early development as synapses are forming. The selective localization of polyribosomes beneath synaptic site and their prominence during periods of synapse growth suggest that they may actually produce proteins for the synaptic junctional region. Although the presence of ribosomes indicates some capabilities for local protein synthesis, the number of ribosomes is low. This suggests that dendrites and the postsynaptic sites that they bear would still require substantial amounts of protein material from the cell body.

Exceptions to the Principle of Dynamic Polarization

For most neurons, the principle of dynamic polarization holds; that is, axons are presynaptic and dendrites are postsynaptic. There are exceptions to the rule, however. In some locations, *axo-axonic* synapses exist, where one class of axon terminal synapses upon another class of terminal that, in turn, synapses with a dendrite; the result is a *synaptic triad,* where there is an axo-axo-dendritic *serial synapse.* For example, the terminals of primary afferents to the spinal cord are contacted by terminals from interneurons. These axo-axonic synapses are thought to mediate *presynaptic inhibition.* By depolarizing the primary afferent terminals that result in *primary afferent depolarization,* these terminals presumably reduce the amount of transmitter released by the primary afferent. The axons of retinal ganglion cells are also involved in synaptic triads in the lateral geniculate nucleus.

Dendrites can also be presynaptic to other dendrites. For example, in the olfactory bulb, granule cells have no axons; instead, they form reciprocal *dendro-dendritic* synapses with the dendrites of mitral cells. Each dendrite possesses presynaptic release sites with accumulations of vesicles that are apposed to postsynaptic membrane specializations on other dendrites. Presynaptic and postsynaptic specializations exist side by side, forming *reciprocal synapses*. The reciprocal synapses form the basis for *local circuit interactions* between neurons without axons and projection neurons. Reciprocal dendro-dendritic synapses also play an important role in the retina.

Neuronal Form and Its Relationship to Function

Despite the enormous length of axons, their exact *form* is not crucial (except in terms of where synapses are made); this is true because axons transmit information in a nondecremental fashion. Once an action potential is initiated in any part of an axon, it will be conducted nondecrementally unless there are some physical discontinuities in the axon or areas where the appropriate ion channels are absent or inactivated. In contrast, most dendrites conduct passively as *cable conductors*. This is important because shape parameters determine cable properties.

Dendritic Form and Integrative Functions

Integration of electrical signals from different synapses depends directly upon the electrotonic properties of the dendrites and the distance between synapses. Both of these features are directly related to the form of the dendrites. Instead of being entirely dependent upon a single "master" input as is true of muscle fibers, neurons receive multiple converging inputs. These inputs are *integrated* by the postsynaptic cell, and the summation of excitation and inhibition is what determines the firing of the postsynaptic cell, which starts at the spike-initiating site, usually the *axon hillock*. The firing depends on a threshold level of depolarization at this site, which is, in turn dependent upon the sum of excitatory and inhibitory input integrated over space and time.

Most synapses operate by opening or closing ion channels, leading to local alterations in current flux. The extent to which a local current at one site affects the membrane potential at another site depends upon the *electrotonic distance* between those sites. The form of the postsynaptic cell determines electrotonic distance; this is because the cable properties of dendrites depend upon the relationship between membrane resistance and internal resistance (Fig. 1.11), and internal resistance depends upon the diameter of the dendrite.

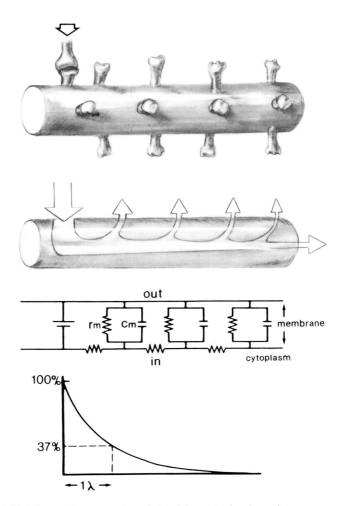

FIGURE 1.11. Electronic properties of dendrites. Activation of a synapse upon a dendrite leads to current fluxes across the dendritic membrane, thus the synapse can be thought of as injecting current locally (*arrow*). The current that flows into the dendrite at the synapse flows along the core of the dendrite and out across the membrane. The current is divided between the cytoplasmic path (the core), and the membrane path depending on the ratio between the internal resistance (Ri) and the membrane resistance (Rm). The membrane has both resistive and capacitive components, and the membrane can thus be schematized as repeating sets of resistors and capacitors in parallel. Because some of the current flows out of the dendrite, the net current flow in the core decreases with distance from the source, and an electronic potential thus decreases with distance as a function of the ratio between Ri and Rm. The characteristic length or length constant (λ) is the distance over which an electrotonic potential falls to 1/e (approximately 0.37) of its initial value. [Reprinted by permission of Elsevier Science Publishing Company from Koester J (1981) Functional consequences of passive electrical properties of the neuron, in Kandel E, Schwartz J (eds): *Principles of Neural Science*. New York, Elsevier. Copyright 1981 by Elsevier Science Publishing Co., Inc.]

Inward current flow at one site upon a dendrite leads to local depolarization at that site. There is a resulting current flow along the core of the dendrite and out across the membrane. At any given site the extent of current flow along these two paths depends upon the ratio between the resistivity of the paths (Fig. 1.11). Essentially, current is continually being divided between the cytoplasmic path and the membrane path, depending on the ratio between internal resistance (R_i) and membrane resistance (R_m). Membrane resistance depends on the properties of the membrane and is thought to be fairly constant at all sites along the cell surface except at synapses. Internal resistance depends on the composition of the cytoplasm, which again is fairly consistent in most dendrites. Assuming that R_m and R_i are reasonably constant across dendrites, the longitudinal resistance of the dendrite varies with the square root of dendritic diameter.

Length Constant

Electrotonic distance is measured as the distance over which an electrotonic potential falls to $1/e$ (approximately 0.37) of its initial value (V_o) (Fig. 1.11). This value is the *characteristic length* or *length constant* of the cable. Because the length constant of a dendrite depends on the ratio between R_m and R_i, and because R_m is constant while R_i varies as a function of the square root of the diameter of the dendrite, the length constant of a dendrite also varies as a function of the square root of the dendritic diameter. As a result, a synapse that is situated at a given physical distance from a spike-initiating zone will actually be at a greater electrotonic distance if it terminates on a small-bore dendrite than if it terminates on a large-bore dendrite.

Shape Index of Excitatory Postsynaptic Potentials

The cellular membrane has both resistive and capacitative components; as a result, the electrical properties of the membrane are schematized as repeating sets of resistors and capacitors in parallel (Fig. 1.11). The membrane capacitance, together with the cytoplasmic resistance, produces a low-pass filter in terms of the current flowing along the long axis of the core. High-frequency components are essentially bled off across the membrane capacitance. As a consequence, the greater the electrotonic distance between a site of current injection and a recording site (the cell body, for example), the slower the intracellular potential (Fig. 1.12). For a given amplitude and duration excitatory postsynaptic potential (EPSP), the depolarization recorded at the level of the cell body will have a longer rise time and lower amplitude the further away the synapse is from the recording site. In this way, the amplitude and duration of a synaptic potential elicited in the dendrite and recorded in the cell body depends on the microanatomy of the neuron.

The relationship between the shape of the EPSP and the site of termination of the active synapse provides a *shape index* for intracellularly

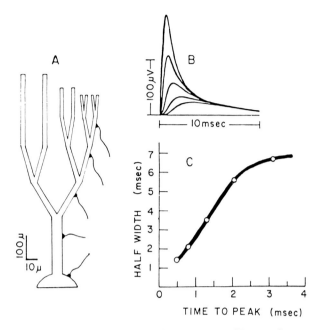

FIGURE 1.12. The effect of electrotonic distance on EPSP waveforms recorded in the cell body. *A* illustrates synapses terminating at different proximodistal locations upon an idealized dendrite: *B* illustrates the expected waveforms and durations of EPSPs generated at these proximodistal locations as would be recorded by an electrode positioned in the cell body. The waveforms were computed based on an electrical model of the branching pattern of the dendrite. Note that the computed waveforms diminish in amplitude and exhibit slower rise times as the synaptic site is moved from the soma distally. *C* illustrates the *shape index* (half width and time to peak) as a function of electrotonic distance from the soma. [From Dodge FA (1979) The nonuniform excitability of central neurons as exemplified by a model of the spinal motoneuron, in Schmitt FO, Worden FG (eds): *The Neurosciences Fourth Study Program*. Cambridge, MIT Press.]

recorded EPSPs, which can be used to compare termination sites of different inputs to a given cell (Fig. 1.12). If it is assumed that all synapses open ion channels for about the same period of time, leading to EPSPs of comparable duration, then synapses on distal dendrites would lead to EPSPs with slow rise times when recorded in the cell body, whereas synapses on proximal dendrites or on the cell body itself would produce EPSPs with fast rise times. Furthermore, if all synapses injected about the same amount of current, leading to EPSPs of constant amplitude, then the depolarization produced at the cell body by distally terminating synapses would also be much less than the depolarization produced by proximally terminating synapses. All synapses are not equal, however, and, to some extent, distally terminating synapses often produce a greater depolarization at the postsynaptic site, thus compensating for the disadvantage that results from the distance between the synapse and the spike-initiating site in the cell body.

Spatiotemporal Summation

The spatial location of synapses also determines the extent of interaction between synapses. If two inputs terminate close to one another, their actions add nonlinearly, that is, the amount of depolarization induced by the two synapses firing together is less than the sum of the two independent EPSPs. This is termed *nonlinear summation* and comes about because the synaptic potential is due to opening or closing of ion channels: If one synapse depolarizes a cell to a level that is near the reversal potential, then opening additional ion channels will have no additional effect on the membrane potential. These kinds of interactions can be important for the processing activities of neurons; indeed, these represent one form of integration.

Dendritic shape parameters permit interactions between synapses within local postsynaptic domains and minimize interactions between synapses that are distant from one another. The spatiotemporal interactions described above only occur between synapses that are near one another (in terms of electrotonic distance). If synapses are distant from one another, then interactions will be minimal. By creating electrotonic distance between certain synapses, dendritic-shape parameters determine the spatiotemporal interactions between different synapses that terminate upon a postsynaptic cell.

Part II
Molecular Anatomy of Neurons, Intracellular Transport, and Gene Expression

Molecular Anatomy of Neurons

As discussed in the previous section, neuronal function depends on an elaborate and highly differentiated form, where different parts of the cell subserve different functions. In neurons, as in other cells, cell form is thought to be regulated by the cytoskeleton. The extremely complex form of neurons implies that neurons may have a cytoskeleton that is different from other less complex cells. Moreover, the construction of different types of structural specializations (axons, dendrites, spines, etc.) is likely to require different types of cytoskeletal elements in different intracellular locations. Both of these predictions have been amply supported by recent studies of the neuronal cytoskeleton.

The Neuronal Cytoskeleton

For any fluid-filled elastic container, the most thermodynamically stable state is a sphere. Departures from spherical shape require either a semirigid membrane (such as for balloons of particular shapes) or internal scaffolding. Cells seem to solve this problem with the latter approach by building a *cytoskeleton,* a semirigid matrix composed of the filamentous structures of the cell (microfilaments, intermediate filaments, and microtubules). The extreme degree of structural differentiation of neurons requires a highly specialized cytoskeleton.

The same three types of filamentous structures exist in neurons as in other cells (Table 1.1). However, there are some special properties of the neuronal cytoskeleton that are presumably related to the complex shape of neurons.

Microfilaments

Microfilaments (about 7 nm in diameter) exist in neurons as in other cells; these are actin filaments. Microfilaments are particularly prominent in the highly motile portions of neurons (growth cones of axons and dendrites) and in dendritic spines. A filamentous lattice consisting of actin filaments is the principal cytoplasmic feature of the head of dendritic spines. The filaments in these structures are actin, as demonstrated by immunocytochemistry using specific antibodies and using heavy meromyosin (S1) fragments to decorate the filaments.

Neurofilaments

In neurons the filaments that are the size of intermediate filaments (10 nm) are termed *neurofilaments.* In general, large axons have a higher neurofilament to microtubule ratio than large dendrites, but both structures contain them. Unlike the filaments that are found in other cell types, which consist of primarily one protein, neurofilaments are composed of a unique

TABLE 1.1. Cytoskeletal Structures.

Type	Size	Composition	Location
Microfilaments	4–5 nm	Actin (mol wt, 43K)	Growth cones and spines
Neurofilaments	10 nm	70K (core) 140K (side arms) 220K (side arms)	Axons and proximal dendrites
Microtubules	23 nm	Tubulin (mol wt, 52 and 56K) MAP (side arms)	Axons and dendrites

MAPs = microtubule-associated proteins

FIGURE 1.13. Neurofilaments in a neuron of the chick spinal cord (Bodian stain). The axon from this neuron (Ax) projects toward the bottom right-hand portion of the figure. The axons of other neurons are visible in the field. Proximal dendrites of the neuron (den) are also stained.

set of three proteins of about 70 kd, 140 kd, and 220 kd (the size depends upon the species). The neurofilaments also differ in morphology from intermediate filaments in other cell types in that neurofilaments possess side arms. These side arms are thought to participate in the interactions between neurofilaments and other elements of the cytoskeleton, particularly microtubules. The core of the filament consists of the 70 kd protein, similar to the intermediate filaments in other cells; the higher molecular weight proteins are thought to form the side arms.

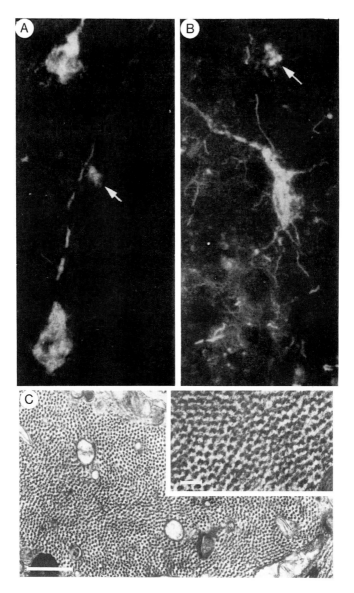

FIGURE 1.14. Neurofilament pathology in Alzheimer's disease. *A, B* Alzheimer's neurofibrillary tangles in the pyramidal cell layer of the hippocampus as revealed by immunocytochemistry using antibodies to brain neurofilament proteins. *C* Electron micrographs of densely packed helical filaments (paired helical filaments) that are found in neurons undergoing degeneration in Alzheimer's disease. Inset shows a high magnification view of an ordered array of paired helical filaments. The calibration bar = 0.1um. [*A* and *B* are from Dahl D, Selkoe DJ, Pero RT et al (1982) Immunostaining of neurofibrillary tangles in Alzheimer's senile dementia with a neurofilament antiserum. *J Neurosci* 2, 113–119. Copyright, Society for Neuroscience. *C* is from Metuzals J, Montpetit V, Clapin DF, et al (1982) Arrays of paired helical filaments in Alzheimer's Disease. *40th Ann Proc Electron Microscop Soc Am* 348–349.]

Neurofilaments are thought to convey structural rigidity to the elements in which they are found. The neurofilaments are most numerous in proximal dendrites and in the axis cylinder, which are the most stable portions of the respective processes. Neurofilaments are not found in the growing tips of axons or in dendritic spines, which are dynamic structures. Neurofilaments are the elements that are stained by the reduced silver methods such as the Bielschosky and Bodian stains, which are important for neuropathological diagnosis. These "silver stains" reveal "neurofibrils" in axons and proximal dendrites that are actually bundles of neurofilaments (Fig. 1.13).

Several degenerative diseases result in abnormalities in neurofilaments that can be revealed with the silver stains. For example, senile dementia and Alzheimer's disease are characterized by the appearance of *neurofibrillary tangles* or *Alzheimer's bodies,* along with neuritic plaques. Neuritic plaques are masses of neuronal processes undergoing degenerative changes. These plaques can be stained with the reduced silver stains or with antibodies against cytoskeletal proteins (Fig. 1.14). Although neurofibrillary tangles are most characteristic of Alzheimer's disease, similar abnormalities are also observed in pathological material from patients with Down's syndrome, postencephalitic parkinsonism, and amyotrophic lateral sclerosis (ALS).

Microtubules

Microtubules in neurons, as in other cells, are composed of tubulin. Both alpha- and beta-tubulin are present, and the tubulin molecules form the wall of the microtubule. Microtubules are found in both axons and dendrites, and it is thought that they are directly responsible for the movement of material in the "rapid phase" of intracellular transport (see below). Recent experiments using high magnification cinematography have revealed that particles are translocated along isolated microtubules in vitro.

Drugs that disrupt microtubules interfere with transport and block the rapid phase entirely. A number of drugs—including colchicine, the "vinca" alkaloyds (vincristine, vinblastine) and colcemid—bind to tubulin and depolymerize microtubules. When neurons are treated with these drugs, transport is disrupted. If the block is applied in a "cuff" around a nerve, organelles accumulate at the site of blockade.

Neuronal microtubules have unique accessory proteins termed microtubule-associated proteins (MAPs). The complement of MAPs is related to the location of the microtubules. Microtubules in dendrites have a high molecular weight MAP (MAP2), which is essentially absent from axons. This protein is also absent from virtually all nonneuronal cells. In constrast, axons have lower molecular weight MAPs in the 50- to 60-kd molecular weight range (tau proteins). The differential labeling of microtubules may be related to the differential routing of material into dendrites versus axons. Other MAPs are found on microtubules in both axons and dendrites.

Axoplasmic Transport and the Distribution of Materials Within Neurons

The high degree of structural specialization poses unique problems for neurons. Not only must these elaborate specializations be constructed in the first place, but they must also be maintained throughout the life of the organism. For this purpose, neurons make use of cellular processes that are, if not entirely unique, at least highly developed in comparison with other cell types. The most important of these cellular processes involves the transport of the material synthesized in the cell body.

As noted above, dendrites have limited capabilities for local synthesis of proteins, and axons have essentially none; thus, axons and dendrites depend on the transport of proteins produced in the cell body. Although the dependence of the axon on the cell body was recognized quite early (and formed the basis of Cajal's statement that the cell body is the trophic center of the neuron), transport was not directly demonstrated until the 1940s in the work of Paul Weiss.

Weiss and his colleagues studied what happened to nerves when they were constricted with ligatures for long periods of time. Such ligatures, if not too tight, constrict the nerve without killing it. In a classic study, Weiss and Hiscoe (1948) showed that constriction resulted in a ballooning of the axon proximal to the constriction and a beading, telescoping, and coiling of the axons. These changes suggested to Weiss that the axoplasm was advancing as a column that "buckled" when it encountered a constriction. When the constriction was removed, the bolus that had accu-

---

▷

FIGURE 1.15. Rapid axoplasmic transport. *A* Axonal transport is evaluated by injecting radiolabeled protein precursors (3H-leucine for example) into regions containing the cell bodies of particular neurons (eg, the dorsal root ganglion or the ventral horn of the spinal cord that contains the cells of origin of motor fibers.) The precursor is incorporated into protein within the neuronal cell bodies and transported down the axon. The rate of transport can be evaluated by harvesting the nerves at different times after the injection, sectioning the nerves into segments, and determining the amount of radiolabeled protein in each segment. A plateau of radioactivity is apparent in the nerve throughout the proximal segment until reaching a crest at the transport front. The transport distance is determined by measuring the distance between the cell bodies and the foot of the front. *B* By sampling at different times after the injection, the front can be seen to progress away from the cell bodies at a defined rate. The scales on the left-hand side refer to the 2-hour and 10-hour samples, and partial scales are illustrated on the right for the 4-, 6-, and 8-hour samples. [*A* is from Ochs S (1977) Axoplasmic transport in peripheral nerve and hypothalamoneurohypophyseal systems, in Porter JC (ed): *Hypothalamic Peptide Hormones and Pituitary Regulation.* New York, Plenum Press. *B* is from Ochs S (1972) Fast transport of materials in mammalian nerve fibers. *Science* 176: 252–254. Copyright 1972 by the AAAS.]

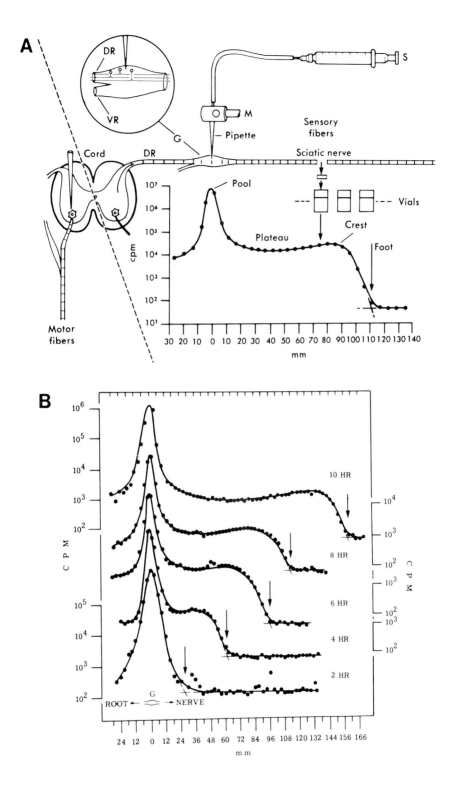

mulated advanced down the axon without dispersing; the rate of advance was about 1 to 2 mm/d. The movement of material was for the most part *somatopedal*, that is, away from the soma. Another term for transport away from the cell body is *orthograde transport*.

A variety of different experimental approaches were used to study transport, but the next major discovery involved the use of radioactive protein tracers. Droz and Leblond (1963) demonstrated that after systemic injection of labeled protein precursors, the label first appeared in the neuronal cell bodies and then appeared as a radioactive front in the axon, which moved at a rate of about 1 to 2 mm/d. This provided an experimental technique that has been used to great advantage to study transport phenomena. Most of the subsequent investigations delivered the tracer directly to the nerve cell bodies and then studied the extent of labeling of nerves at various times after the tracer pulse (Fig. 1.15).

In addition to providing an experimental technique for studying transport, these investigations also revealed that there were actually multiple transport systems that delivered different types of material and that were destined for different intracellular locations.

"Slow" Axonal Transport

Most of the material transported into the axon moves in the so-called "slow component" of axoplasmic transport. This is the component that moves at a rate of about 1 to 2 mm/d, which was described originally by Weiss. Thus, material synthesized in the cell body may take days or weeks to transit the axon. Most of the structural proteins of the axon and the cytoskeleton move via slow transport; however, different proteins of the cytoskeleton move at slightly different rates, so that the slow component actually consists of subcomponents that transport different material. Actin and neurofilament proteins move via slow component B (SCb), whereas tubulin moves via slow component A (SCa). The difference between the rates of SCb and SCa is minor, however, compared with the difference between slow transport and rapid transport (see below).

Rapid Transport

There is also a rapid form of transport that moves material down the axon at a rate of up to 400 mm/d. Rapid transport was discovered only because of the power of the radioactive tracer techniques (Fig. 1.15). A small amount of the total radioactivity that is incorporated into protein in the nerve cell body moves down the axon at a much faster rate than other proteins. This rapid transport is highly dependent on oxidative metabolism within the nerve, since agents that interfere with oxidative metabolism lead to a rapid disruption of fast transport.

The material carried by rapid transport is mainly "particulate"; that is, it is associated with membranous organelles rather than being part of the cytoskeleton or one of the soluble proteins of the cytoplasm. Most of the material transported in this phase is probably associated with small vesicular structures that can be seen in the axon. This rapid transport is thought to be mediated by microtubules, since the rapid phase of transport is blocked by microtubule poisons, such as colchicine.

Fast transport delivers material preferentially to the synaptic terminals. Whereas the rate of accumulation of radioactive material in the axon proper or in preterminal arborizations is measured in days following delivery of radioactive protein precursors to the cell body, labeled material accumulates in the nerve endings within hours. Biochemical studies reveal that many of the proteins that travel in the rapid phase of transport are membrane glycoproteins.

Retrograde Transport

In addition to the orthograde transport of material from the cell body into the processes, there is also a *retrograde transport* of material from the terminals to the cell body. This retrograde transport phenomenon was initially characterized by injecting the protein tracer horseradish peroxidase into the terminal zones of particular neurons. The protein is taken up by the terminals, encapsulated within vesicles, and transported back to the cell body, where the vesicles combine with lysosomes. Virtually any protein molecule can be taken up and transported in a retrograde direction. In addition, fluorescent dyes that bind to proteins are readily transported, presumably in conjection with the proteins to which the dyes bind. Both of these properties have been used to advantage for tract-tracing studies, providing an experimental technique to identify the cells of origin of projections to particular sites.

Retrograde transport provides a mechanism for the cell body to sample the environment around its synaptic terminals. For some neurons, the maintenance of synaptic interconnections depends on "trophic interactions" between presynaptic neurons and their targets. There is growing evidence that there is an exchange of material across the synapse. The best example of a retrograde trophic substance is *nerve growth factor (NGF)*. As will be discussed in more detail in Chapter 5, NGF is produced in the targets of NGF-sensitive fibers (for example, the end organs of the sympathetic nervous system). The NGF is released from these targets and taken up by the presynaptic neurons, where it is transported back to the cell body. In the sympathetic nervous system, where NGF was discovered, the factor stimulates the growth of sympathetic axons into the target area and sustains the sympathetic neurons after the formation of connections.

In addition to target-derived factors, there is evidence for an exchange of material from the presynaptic terminal to the target. The transneuronal transport of labeled material in the visual system (which has been used to label the geniculo-cortical projections associated with one eye) takes advantage of this *transneuronal transport*.

Pathologies of Axoplasmic Transport

A number of compounds that lead to peripheral neuropathies as a result of the disruption of axonal transport have been discovered. For example, the neurotoxic effects of *n*-hexane or 2-hexanone were discovered in individuals who had been chronically exposed to these compounds. These compounds are used extensively as intermediates in the synthesis of perfumes, rubber accelerators, dyes, and wood stains and are also found as gasoline additives and tanning agents. Systemic administration leads to slowly progressive muscle weakness attributable to distal axonal degeneration.

Early signs of neurotoxicity involve the disruption of axonal transport. The affected axons exhibit large neurofilament-filled swellings with a paucity of neurofilaments distal to the enlargements. It is believed that the compounds impair the movement of the neurofilaments, which leads to a degeneration of the peripheral portion of the nerve.

Acrylamide, *p*-bromophenylacetylurea, zinc pyridinethione, and B,B′ iminodipropionitrile (IDPN) also have neurotoxic effects that seem to be related to transport blockade. Most of these were initially discovered as a result of the appearance of peripheral neuropathies in individuals who had been exposed to the substances. In addition, chronic treatment with disulfiram (Antabuse) is associated with peripheral neuropathy, which is correlated with a disruption of transport. Vitamin E deficiency and diabetes are also associated with peripheral neuropathies that are thought to involve disrupted transport. All of these peripheral neuropathies eventually lead to "dying back" of the peripheral process of the affected axon as a consequence of the disruption of the delivery of material from the cell body.

A blockade of axonal transport is also evident in a number of retinopathies and optic neuropathies. For example, since rapid transport depends on oxidative metabolism, occluded arteries supplying the nerve fiber layer of the eye lead to transport blockade that results in the same kinds of swellings that occur following nerve constriction. Increased intraocular pressure also leads to transport blockade as a result of compression of the nerve fiber bundles at the lamina cribrosa. Papilledema also results in axonal swelling in the anterior part of the optic nerve head. The axonal changes induced by elevated intraocular pressure or papilledema can be reversed if the causative factors are eliminated. If the transport blockade persists, however, degeneration of the distal axons can occur, leading to permanent visual impairment.

Retrograde transport may be one of the mechanisms for the entry of viruses into the CNS. It has long been known, for example, that the rabies virus appears to spread along peripheral nerves to the CNS from the inoculation site at the periphery. It is thought that the herpes simplex virus also reaches neurons as a result of retrograde transport. It is transported to neurons in the trigeminal nerve and other sites in the CNS after injection into the snout. Herpes simplex virus remains latent in trigeminal ganglia between recurrences of the herpetic lesions, and when activated, appears to travel along the peripheral branches of axons to the lip or conjunctiva.

In the same way that transport mechanisms provide a way for viruses to enter the nervous system, they may eventually provide the most direct means to treat the infections. For example, topical application of antiviral agents to the peripheral fields of the affected axons may lead to uptake, retrograde transport, and delivery of the agents to the affected cells in a highly specific fashion.

Dendritic Transport

Dendrites also possess transport mechanisms that permit the delivery of material from the cell body. This is not surprising since dendrites, like axons, extend for considerable distances from the neuronal cell body, and the materials comprising the dendrites must reach the distant sites via some mechanism. Interestingly, the transport machinery in dendrites appears to convey different materials than that of axons.

For example, studies of neurons grown in tissue culture have revealed that there is a selective dendritic transport of RNA. When neurons in culture are pulse-labeled with 3H-uridine, an RNA precursor, the precursor is incorporated into RNA. When these neurons are fixed and prepared for autoradiography in such a way that soluble RNAs (transfer RNA) are washed out, the recently synthesized (labeled) RNA is initially present only in the nucleus. Over time, the recently synthesized RNA appears in the neuronal cytoplasm and then in the dendrites (Fig. 1.16). At no time does label appear in the axons, however. Thus, the form of RNA that is detected by these autoradiographic procedures (presumably ribosomal and messenger RNA, but not transfer RNA) appears to be transported selectively into dendrites and not into axons.

The rate of this dendritic transport of RNA can be estimated by measuring the average distance that the label migrates into dendrites at various times after the pulse. Such calculations reveal that RNA moves at about 0.4 mm/d—a rate that is similar to some forms of slow axonal transport. Like the rapid form of axonal transport, dendritic transport of RNA seems to be energy dependent, since it is blocked by metabolic poisons such as dinitrophenol (DNP) or sodium azide.

The differences between axons and dendrites in the materials that are conveyed by the transport machinery are of considerable interest, since it seems almost certain that different materials would need to be transported into axons and dendrites because of their different structure and function. Many questions remain to be answered about dendritic transport,

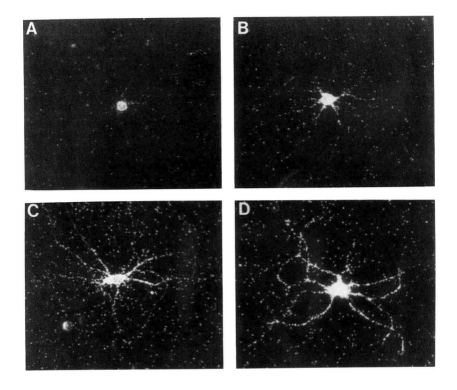

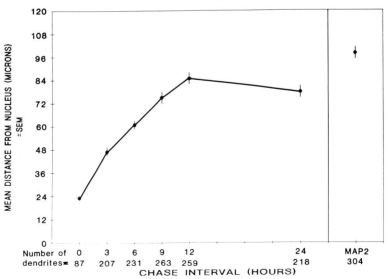

E

DISTANCE OF MIGRATION OF 3H−RNA IN DENDRITES
25 NEURONS

Number of	0	3	6	9	12		24	MAP2
dendrites=	87	207	231	263	259		218	304

CHASE INTERVAL (HOURS)

including why the dendritic transport machinery conveys different materials than the transport machinery of axons, whether materials other than RNA are transported selectively, and whether there are multiple rates of transport.

Gene Expression in Neurons

The basic mechanisms of gene expression in neurons are similar to those that operate in other eucaryotic cells. Thus, proteins are encoded by genes that are located on chromosomes in the cell nucleus. The genes possess coding regions *(exons)* that are interrupted by noncoding regions *(introns)*, and the genes are transcribed by the enzyme RNA polymerase II, which generates a single stranded *heterogeneous nuclear RNA* (hnRNA) that is a complete copy of the gene sequence (complete with noncoding introns). A 5'-cap structure is added to the hnRNA soon after transcription is initiated, and 100-200 adenylate residues are added posttranscriptionally to the 3'-end of most gene transcripts to form a *poly(A) tail*. The hnRNA molecule is cut so as to remove the noncoding introns, and the exons are spliced to form a completed mRNA molecule. Different mRNAs can be produced from the same hnRNA by *alternative RNA splicing*, thus providing another mechanism for generating protein diversity. Alternative splicing is particularly important for generating related proteins with minor sequence differences. For example, in neurons, alternative splicing is known to play a role in generating different types of neuropeptide neurotransmitters, and is probably also important for numerous other proteins.

The processing of the hnRNA into the completed mRNA molecule takes place in the nucleus, and the mRNA is then translocated from the nucleus to the cytoplasm through a process of *nucleocytoplasmic transport*. Once

---

FIGURE 1.16. Dendritic transport of RNA. Dendritic transport was evaluated by pulse-labeling neurons in culture with an RNA precursor (3H uridine). RNA is synthesized only in the cell nucleus; thus, immediately after the pulse, labeled RNA is present only in the nucleus. The neurons were fixed and prepared for autoradiography at various times after the pulse to evaluate the distribution of recently synthesized (labeled) RNA. *A* Distribution of recently synthesized RNA at the end of the labeling period. Note that grains are found almost exclusively over the nucleus. *B-D* Distribution of recently synthesized RNA at 3, 6, and 24 hours after the pulse. Note the progressive migration of label into dendrites. *E* Time course of dendritic transport. The rate of transport is determined by measuring the distance between the center of the nucleus and the furthest extent of labeling within each dendrite. The average length of the dendrites themselves was determined by measuring the length of dendrites in cultures stained for the microtubule-associated protein, MAP2, which is a selective marker for dendrites. [Reprinted by permission from *Nature* 330: 477–479. Copyright © 1987 Macmillan Magazines Ltd.]

in the cytoplasm, the mRNA is translated on ribosomes, where a protein is synthesized according to the code provided by the mRNA. The mRNA molecule also contains "untranslated" regions at the 3' and 5' ends which may be important for determining the rate of translation of particular mRNA molecules, the stability (half-life) of the mRNA, and possibly may determine the intracytoplasmic transport of the mRNA. Primary molecular biology texts should be consulted for more details about the general mechanisms of regulation of gene expression in eucaryotic cells.

While the basic mechanisms of gene expression in neurons are similar to those of other eucaryotic cells, certain aspects of neuronal gene expression may be somewhat unique. First, there is evidence to suggest that much of the regulation of gene expression in brain occurs posttranscriptionally, at the stage of mRNA processing. Thus, the levels of particular mRNA's in the neuronal cytoplasm may be regulated independently of the levels of the primary gene transcripts (the hnRNAs). Second, there is evidence, although presently somewhat controversial, that a large number of mRNA's in brain are *not* polyadenylated, a situation which is not found in other cell types. These non-polyadenylated mRNA's may represent a special class of mRNA that is especially important to neuronal function. Finally, the number of genes expressed in the nervous system is much higher than in other tissues.

Brain-Specific and Cell Type-Specific mRNAs

Given the elaborate form of neurons, and their intricate connections, it is certain that neuronal function depends upon the expression of a large number of genes. Similarly, given the many different "phenotypes" of neurons, it is certain that different genes are expressed in different neuron types. Indeed, based upon analyses of mRNA heterogeneity, it has been estimated that about ⅓ of the mammalian genome is dedicated exclusively to neuronal function. That is, about ⅓ of the mammalian genome is actively transcribed in brain, whereas the amount of the genome that is transcribed in non-neuronal tissues such as liver or kidney is about two- to three-fold lower. In addition, it has been estimated that about half of the mRNA molecules that are present in brain are brain-specific; most of the mRNA that is specific to brain is present in relatively low abundance, suggesting that it may be expressed in subsets of neurons or glia.

Certainly, some of the differences between neurons in the genes that are expressed are related to the types of neurotransmitters used, and the types of neurotransmitter receptors that are expressed. These issues are discussed in more detail in Chapters 3 and 4. However, there are undoubtedly a host of other genes that are "neuron-specific", coding for proteins that are crucial or the development of the specialized processes of neurons (axons and dendrites) and their mosaic surface membranes. Moreover, there are unquestionably large differences *between* neurons in their structural features, membrane properties, and types of synaptic connections that would require type-specific proteins.

Neuron-Specific Proteins

Early biochemical studies of the types of proteins that were present in brain revealed a few proteins that were either unique or at least unusually abundant in brain. Examples include the glial protein *S100*, so named because of its solubility in 100% ammonium sulfate, and *neuron-specific enolase*. However, the biochemical studies could only reveal abundant proteins, and most abundant proteins are also found in non-neuronal cells. Many of the most interesting proteins in brain are present in relatively low abundance, being expressed by only a small percentage of the total number of neurons. Except for the molecules involved in neurotransmission (transmitters themselves and receptors), the properties and cellular localization of these presumptive neuron-specific proteins are unknown. As a result, there until recently was no useful experimental strategy that permitted a search for proteins of low abundance that were unique to particular cell types. Recently, thanks to the development of techniques to study molecules at the level of individual cells such as monoclonal antibody technology, and gene cloning techniques it has become possible to characterized molecules that are unique to particular types of neurons that may be related to the differentiation of neuronal phenotypes.

A number of neuron specific proteins have now been identified as a result of the application of antibody technology. For example, a number of neuron-specific cytoskeletal proteins have been identified in this way. Thus, MAP2, the neurofilament proteins, and the Tau proteins are all for the most part neuron-specific. Moreover, certain tubulin isotypes are unique to neurons.

Use of Antibodies to Identify Proteins Unique to Particular Classes of Neurons

Not surprisingly, proteins that are involved in neurotransmission are specific to particular classes of neurons. In addition, certain enzymes that are involved in second messenger systems are specific to particular classes of neurons. For example, a cyclic guanosine monophosphate (cGMP)-dependent protein kinase is highly concentrated in cerebellar Purkinje cells, and is essentially undetectable in most other cell types. These aspects of molecular differentiation of neurons will be discussed further in Chapters 3 and 4.

Antibody technology has also provided a means to identify region-specific or cell-specific proteins even before actually identifying the proteins. One strategy involves the production of antibodies to particular brain regions by using dissected regions for immunization. The antibodies gen-

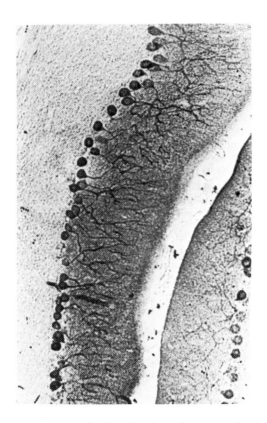

FIGURE 1.17. Immunocytochemical localization of a protein that is specific to cerebellar Purkinje cells (termed cerebellin). Antibodies were prepared against cerebellar tissue, and used for immunocytochemistry. The antigen is localized specifically in Purkinje cells. [From Slemmon JR, Danho W, Hempstead JL, Morgan JI, Cerebellin: A quantifiable marker for Purkinje cell maturation. *Proc Nat Acad Sci* 82: 7145–7148.]

erated in this way can then be screened by immunocytochemistry to determine whether particular cell types are labeled. In this way, proteins specific to cerebellar Purkinje cells were identified (Fig. 1.17).

Another strategy involves the production of panels of monoclonal antibodies against crude neuronal protein preparations. The monoclonal antibodies can then be screened to find antibodies that are expressed in different neuronal types. This approach has revealed a variety of proteins that are unique to single cell types or small subsets of cells in the nervous system. Some of these antigens are cytoplasmic, suggesting a difference in the biochemical activities of the neuron. Other unique antigens are membrane-associated, and may reflect a molecular differentiation of the membranes of different types of cells.

In general, the functional significance of many of the proteins that have been identified through the production of panels of monoclonal antibodies is not known. All that is known for many is that the proteins are specific to particular neuron types. However, by using monoclonal antibodies to isolate and purify the protein (by immunoaffinity), and sequencing the purified protein, it is then possible to construct complementary oligonucleotide probes. These oligonucleotide probes can then be used to identify the genes encoding the protein, to identify the chromosomes which carry the genes (by hybridization analysis), and to identify the sites of synthesis of the protein by ''Northern'' blot hybridization or *in situ* hybridization in tissue.

Use of In Situ Hybridization to Evaluate Gene Expression in Individual Neurons and Populations of Neurons

A second important approach that has provided information about differences in gene expression in different types of neurons involves *in situ* hybridization using probes that are specific for particular mRNA's. The retroviral enzyme *reverse transcriptase* can be used for copying mRNA molecules into complementary DNA (cDNA), which is quite stable. These cDNA's retain the ability to hybridize specifically to the complementary mRNA molecules; thus, the cDNA's can be labeled, and hybridized to mRNA that is present in cells *in situ*. Complementary RNA probes can also be synthesized from the cDNA, and these ''ribo probes'' have several technical advantages over cDNA probes for *in situ* hybridization. *Antisense* probes that are complementary to the mRNA molecule can be synthesized, labeled, and used for *in situ* hybridization in the same way as cDNA probes. *Sense* probes which have the same nucleotide sequence as the mRNA can be used as controls (since these are not complementary to a mRNA sequence, their binding should not reflect specific hybridization). A more recent strategy has been to synthesize *oligonucleotide probes* based on the known protein sequence or the known nucleotide sequence of the mRNA.

In situ hybridization has been used with considerable success to localize the mRNA's for peptide neurotransmitters, thus identifying neurons that produce the neuropeptides, and identifying the intracellular sites of neuropeptide biosynthesis (Fig. 1.18). Similar approaches are being used with mRNAs for many other molecules, to determine whether particular messages are uniquely localized in particular cell types.

Taken together, the application of the techniques of cellular and molecular biology have already revealed a number of neuron-specific proteins that include components of the cytoplasm and cytoskeleton, membrane proteins, neurotransmitter enzymes, neurotransmitter receptors, ion channels including the sodium channel, signal transducers (rhodopsin and opsin in the retina), enzymes involved in second messenger systems, and

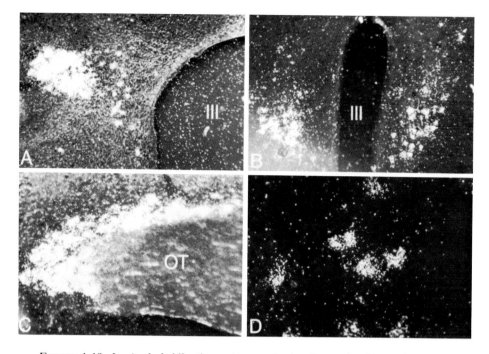

FIGURE 1.18. *In situ* hybridization using synthetic oligonucleotide probes to the neuropeptide arginine-vasopressin (AVP). Oligonucleotides complementary to the AVP mRNA sequence were synthesized and labeled with [125]I. The probes were then hybridized to tissue sections. *A-D* illustrate dark field photomicrographs of autoradiographic preparations prepared following *in situ* hybridization. Silver grains appear as bright spots. Note the heavy labeling of neurons in specific nuclei. *A, B* Magnocellular neurons of the paraventricular nucleus (III = third ventricle); *C* neurons of the supraoptic nucleus (OT = optic tract); *D* higher magnification view of individual neurons in the paraventricular nucleus. [From Lewis ME, Arentzen R, Baldino, F Rapid, high-resolution *in situ* hybridization histochemistry with radioiodinated synthetic oligonucleotides. *J Neurosci Res* 16: 117–124.]

growth factors. Many of these will be discussed in later chapters. Despite rapid progress, the application of the tools of cellular and molecular biology to the nervous system is in its infancy; new strategies will undoubtedly evolve, and are likely to dramatically increase our understanding of neuronal function at the molecular level.

Supplemental Reading

General

Jacobson M (1978) *Developmental Neurobiology*. New York, Plenum Press.

Kuffler SW, Nicholls JG (1976) *From Neuron to Brain*. Sunderland, MA, Sinauer Assoc Inc

Purves D, Lichtman JW (1985) "Principles of Neural Development." Sunderland, MA, Sinauer Assoc Inc

McGeer PL, Eccles JC, McGeer EG (1987) *Molecular Neurobiology of the Mammalian Brain*. New York, Plenum Press

The Cytology of Neurons

Ramon y Cajal S (1911) *Histologie du Système Nerveux de l'Homme et des Vertébrés* (2 vols). (Axoulay L, trans) Reprinted by Instituto Ramon y Cajal del CSIC, Madrid, 1952–1955

Peters A, Palay SL, Webster DeF (1976) *The Fine Structure of the Nervous System: The Neurons and Supporting Cells*. Philadelphia, WB Saunders Co

Dendritic Form

Hillman DE (1979) Neuronal shape parameters and substructures as a basis of neuronal form, in Schmitt FO, Worden FG (eds): *The Neurosciences Fourth Study Program*. Cambridge MA, MIT Press, pp 477–498

Ramon-Moliner E (1975) Specialized and Generalized Dendritic Patterns, in Santini M (ed): *Golgi Centennial Symposium, Proceedings*. New York, Raven Press, pp 87–100

Cytoskeleton

Alberts B, Bray D, Lewis J et al. (1983) *Molecular Biology of the Cell*. New York, Garland

Caceres A, Binder LI, Payne MR et al. (1983) Differential subcellular localization of tubulin and the microtubule associated protein MAP2 in brain tissue as revealed by immunocytochemistry. *J Neurosci* 4:394–410

Lasek RJ, Shelanski ML, Brinkley BR et al. (1981) Cytoskeletons and the Architecture of Nervous Systems. *Neurosci Res Program Bull, 19*. Cambridge MA, MIT Press

Olmsted JB (1986) Microtubule-associated proteins. *Ann Rev Cell Biol* 2:421–57

Axonal Transport

Droz B, Leblond CP (1963) Axonal migration of proteins in the central nervous system and peripheral nerves as shown by radioautography. *J Comp Neurol* 121:325–346

Grafstein B (1977) Axonal transport: the intracellular traffic of the neuron, in Brookhart JM, Mountcastle VB (section eds): *Handbook of Physiology*, Section 1: *The Nervous System*. Baltimore, Williams and Williams, pp 691–717.

Ochs S (1972) Fast transport of materials in mammalian nerve fibers. *Science* 176:252–260

Lasek RJ, Garner JA, Brady ST (1984) Axonal transport of the cytoplasmic matrix. *J Cell Biol* 99:212–221

Weiss P, Hiscoe H (1948) Experiments on the mechanism of nerve growth. *J Exp Zool* 107:315–395

Retrograde Transport

Brunso-Bechtold JK, Hamburger V (1979) Retrograde transport of nerve growth factor in chicken embryo. *Proc Natl Acad Sci USA* 76:1494–1496

LaVail JH, LaVail MM (1972) Retrograde axonal transport in the central nervous system. *Science* 176:1416–1417

Thoenen H, Otten U, Schwab M (1979) Orthograde and retrograde signals for the regulation of gene expression: the peripheral sympathetic nervous system as a model, in Schmitt FO, Worden FG (eds): *The Neurosciences, Fourth Study Program.* Cambridge MA, MIT Press, pp 911–928

Neuronal Form and Its Relation to Function

Nicholson C (1979) The nonuniform excitability of central neurons as exemplified by a model of the spinal motoneuron, in Schmitt FO, Worden FG (eds): *The Neurosciences, Fourth Study Program.* Cambridge MA, MIT Press, pp 477–498

Rall W (1977) Core conductor theory and cable properties of neurons, in Brookhart JM, Mountcastle VB (section eds): *Handbook of Physiology, Section 1: The Nervous System.* Baltimore, Williams and Williams, pp 39–97

Gene Expression in Neurons

Bantle JA, Hahn WE (1976) Complexity and characterization of polyadenylated RNA in mouse brain. *Cell* 8:139–50

Beckman SLM, Chikaraishi DM, Deeb SS et al. (1981) Sequence complexity of nuclear and cytoplasmic RNAs from clonal neurotumor cell lines and brain sections of the rat. *Biochemistry* 20: 2684–92

Chikaraishi DM (1979) Complexity of cytoplasmic polyadenylated and non-adenylated rat brain ribonucleic acids. *Biochemistry* 18:3250–56

Douglass J, Civelli O, Herbert E (1984) Polyprotein gene expression: Generation of diversity of neuroendocrine peptides. *Ann Rev Biochem* 53:665–715

Kaplan BB, Finch CE (1982) The sequence complexity of brain ribonucleic acids, in Brown IR (ed): *Molecular Approaches to Neurobiology.* New York, Academic Press, pp 71–98

Leff SE, Rosenfeld MG, Evans RM (1986) Complex transcriptional units: Diversity in gene expression by alternative RNA processing. *Ann Rev Biochem* 55:1091–2117

Lewis SA, Villasante A, Sherline P et al. (1985) Brain specific expression of MAP 2 detected using a cloned cDNA probe. *J Cell Biol* 102:2098–2105

McCarthy MP, Earnest JP, Young EF et al. (1986) The molecular neurobiology of the acetylcholine receptor. *Ann Rev Neurosci* 9:383–413

Milner RJ, Sutcliffe JG (1983) Gene expression in rat brain. *Nucl Acids Res* 11:5497–5520

Nawa H, Kotani H, Nakanishi S (1984a) Tissue-specific generation of two pre-protachykinin mRNAs from one gene by alternative RNA splicing. *Nature* 312:729–34

Sutcliffe JG (1988) mRNA in the mammalian central nervous system. *Ann Rev Neurosci* 11:157–98

CHAPTER 2

Cell Biology of Glia

Introduction

Despite the preeminent role that neurons play in brain function, glial cells are considerably more numerous. In fact, glia outnumber neurons by about 10:1 and make up about half the bulk of the nervous system. Glial cells were first recognized as a distinct cell type in the mid 1800s. The name "neuroglia," coined by Rudolph Virchow, means "nerve glue," since these cells were thought to play a supportive role in brain function, essentially holding neurons in their proper positions. Glial cells are still thought to be *satellite cells* to neurons or helper cells; nevertheless, their role is considerably more complex than these terms imply. Indeed, neurons and glia are now thought to interact in highly interdependent ways, influencing each other's development, differentiation, and physiological function. This chapter describes the different types of glia in the central and peripheral nervous system and discusses some of the functional roles that these "satellite" cells play.

Glial Cells of the CNS

In the CNS there are two principal types of glia that can be distinguished by size and embryonic origin: *macroglia,* including astroglia and oligodendroglia, are the larger types of glial cells, originating from the neural plate. *Microglia* are smaller and are thought to originate from the mesoderm.

Astroglia or *astrocytes* have small cell bodies, but possess extensive and highly branched processes. There are two kinds of astrocytes, which differ in appearance and location (Figs. 2.1 and 2.2). *Fibrous astrocytes* are found in the white matter. They are classified as fibrous because they have large numbers of *glial filaments,* which are a special type of intermediate filament (Fig. 2.3A). Intermediate filaments, or 10-nm filaments, are found in a variety of cell types, but in astrocytes the intermediate

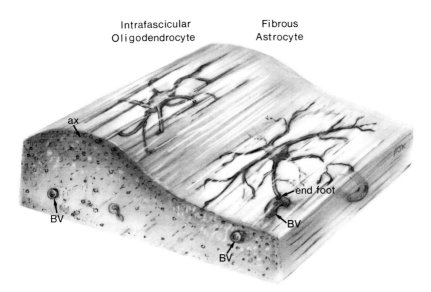

Intrafascicular
Oligodendrocyte

Fibrous
Astrocyte

FIGURE 2.1. Types of glial cells found in white matter. Figure 2.12 illustrates the way in which oligodendrocyte processes give rise to myelin sheaths around axons. ax = axons; BV = blood vessel.

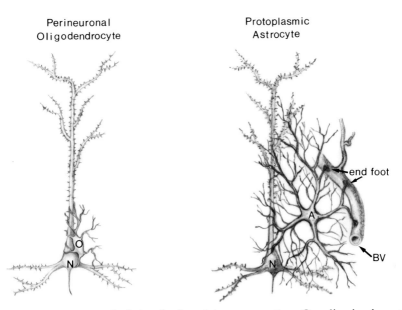

Perineuronal
Oligodendrocyte

Protoplasmic
Astrocyte

FIGURE 2.2. Types of glial cells found in gray matter. O = oligodendrocyte; N = neuron; A = astrocyte, BV = blood vessel.

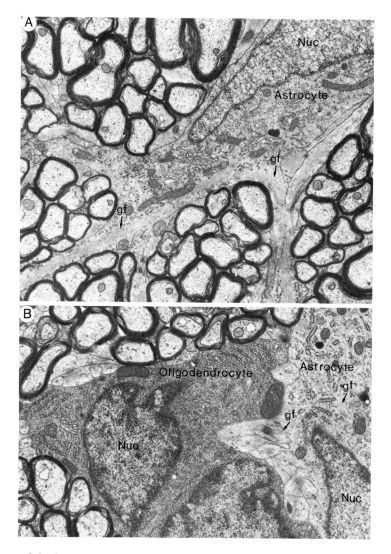

FIGURE 2.3. Astrocytes and oligodendrocytes in a myelinated fiber tract (optic nerve). *A* Cell body and processes of an astrocyte. gf = glial filaments. *B* Cell body of an oligodendrocyte. Note the dark cytoplasm and the expanded cisternae of the Golgi apparatus. The cell body of an astrocyte is visible on the right. (Courtesy of Dr. P. Trimmer.)

filaments consist of a protein that is unique to glia—*glial fibrillary acidic protein (GFAP)*. Fibrous astrocytes give rise to extensive and highly branched processes that intertwine between bundles of axons in the white matter (Fig. 2.3A).

Protoplasmic astrocytes are similar in overall form except that they have fewer glial filaments. They are abundant in gray matter around nerve cell bodies, dendrites, and synapses. The processes of protoplasmic astrocytes insinuate themselves into the spaces between neurons and form thin "vellate" (veil-like) processes around neuronal elements, including cell bodies, axons, dendrites, and synapses (Fig. 2.4).

Both types of astrocytes form *"glial end feet"* on blood vessels. The glial end foot is bounded by basal lamina, and thus the glial end foot can be thought of as forming the boundary between CNS tissue and vascular elements (Fig. 2.5). In this way, glial end feet form a *perivascular glial limiting membrane or glia limitans* similar to that found at the surfaces of the brain (see below). The anatomical relationship between astrocytic end feet and blood vessels has led to speculations that the glial processes might represent the substrate of the blood-brain barrier. However studies in the late 1960s by M. Brightman and T. Reese convincingly showed that

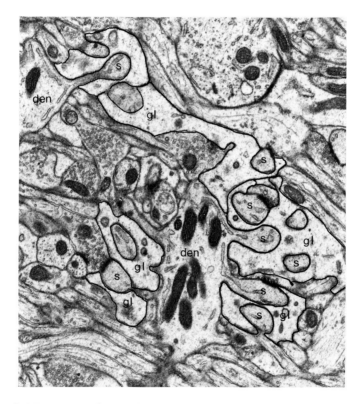

FIGURE 2.4. Processes of protoplasmic astrocytes surrounding synaptic terminals and dendrites in the neuropil (rat cerebellum). The membrane of the astrocytic processes is outlined. Note that the astrocyte surrounds the dendrites and their spines. gl = glial cytoplasm; s = spine; den = dendrite of Purkinje cell.

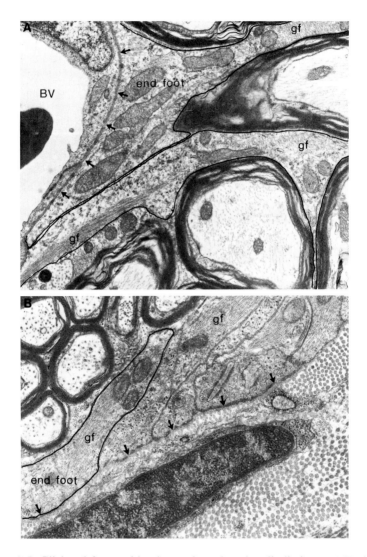

FIGURE 2.5. Glial end feet on blood vessels and at the glia limitans. *A* End foot of astrocytes on a cerebral blood vessel. Membrane of one end foot is outlined. *B* End feet of astrocytes at the glia limitans (rat optic nerve). Membrane of one end foot is outlined. (*Arrows* indicate the basal lamina). gf = glial filaments, BV = blood vessel. (*B* courtesy of Dr. P. Trimmer.)

large molecules such as horseradish peroxidase readily diffused past the glial end feet when introduced into the ventricular system. Instead, diffusion was limited by the tight junctions between endothelial cells. Thus, it is now thought that the blood-brain barrier is formed primarily by endothelial cells, with the glial end feet subserving some other function (see below).

Astrocytes also form the *glia limitans* at the surface of the brain (Fig. 2.6). This continuous sheet of glial processes and cell bodies arises partly from astrocytes at the surface and partly from deeply situated astrocytes that send their processes toward the surface. The external surface of the glia limitans is again bounded by basal lamina, which is situated between the glia and the *pia mater*.

Astrocytes can be easily identified at either the light or the electron microscopic level because their cytoplasm contains relatively few organelles and thus appears lighter than the surrounding neuropil. This is particularly true in gray matter, where astrocytic processes appear quite lucent in comparison to neuronal processes (Fig. 2.4). In addition, glial cells often contain glycogen, which is not usually seen in neurons. The overall morphology of astrocytes is very complex and can best be appreciated using stains that highlight the processes of individual cells within surrounding tissue. Fortunately, like neurons, astrocytes sometimes stain with the Golgi method; this property provided early anatomists a means to define astrocytic form. The recent discovery of specific molecular markers (GFAP) now permits a reliable identification using immunocytochemistry (Fig. 2.7).

Oligodendroglia or *oligodendrocytes* are smaller than astrocytes and have fewer and shorter processes. The oligodendrocytes are, like the astrocytes, of two types (Figs. 2.1 and 2.2). *Intrafascicular oligodendroglia* are found in fiber tracts where they are the myelin-forming elements (Fig. 2.3). *Perineuronal oligodendroglia* are found near neuronal cell bodies as satellite cells (Fig. 2.8). The perineuronal oligodendrocytes come into close contact with neuronal cell bodies and dendrites. Oligodendrocytes can be identified microscopically by their dark-staining cytoplasm (Figs. 2.3 and 2.8). Like the astrocytes, they also sometimes stain with Golgi techniques.

Microglia are the smallest of the glial elements. They have short, thin processes that make contact with neurons and capillaries. Microglia have ultrastructural features that are identical with those of connective tissue

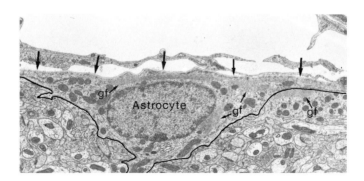

FIGURE 2.6. Astrocytes forming the glia limitans at the surface of the cortex (rat cerebral cortex). *Arrows* indicate the basal lamina; gf = glial filaments. The membrane of the astrocyte is outlined.

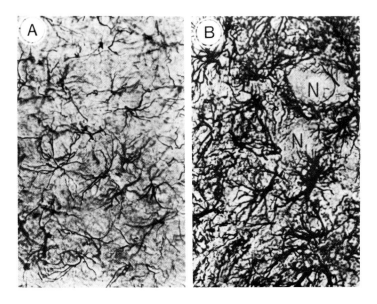

FIGURE 2.7. Staining of astrocytes using antibodies to glial fibrillary acidic protein (GFAP). *A* Astrocytes in the normal spinal cord. Glial filaments in astrocytes consist of a unique protein termed GFAP. Thus, immunocytochemical techniques can be used to identify astrocytes in the CNS. *B* Reactive astrocytes in the injured spinal cord. Astrocytes undergo hypertrophy in response to injury, and part of this reaction response involves an increase in GFAP. Note the envelopment of two neurons (N) by the astrocytic processes. (X 360) [From Reier PJ (1986) Gliosis following CNS injury: the anatomy of astrocytic scars and their influences on axonal elongation, in Federoff S, Vernadakis A (eds): *Astrocytes: Cell Biology and Pathology of Astrocytes*. Academic Press, NY, pp 263–324.]

macrophages, and they are thought to have similar roles. The microglia, unlike the two types of macroglia, can migrate even in the mature CNS, and they play an important function in the removal of degeneration debris following injury (see below).

Ependymal cells line the ventricles as an epithelial layer one cell thick. The cells contact one another through gap junctions, and their ventricular faces have numerous microvilli and cilia. The motile cilia are thought to facilitate the movement of cerebrospinal fluid. During early development, the cells often have a basal process that extends into the substance of the CNS. In the spinal cord and other locations with a central canal, these processes are radially oriented. Such cells are termed *tanycytes, ependymal astrocytes, or ependymoglial cells*. The basal process of these cells is usually withdrawn during development. Nevertheless, during early development, these radial glia are crucial for the migration of neurons and for guiding the growth of axons (see Chapters 5 and 7).

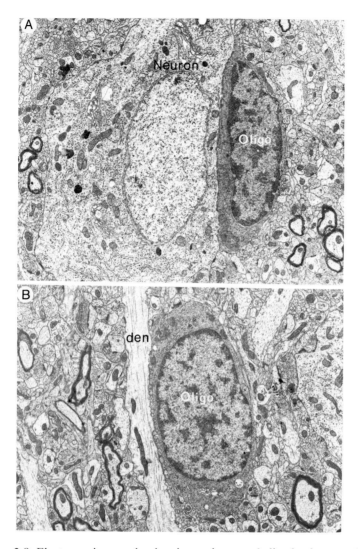

FIGURE 2.8. Electron micrographs showing perineuronal oligodendrocytes (rat ce-rebral cortex). *A* Perineuronal oligodendrocyte in close apposition to a neuronal cell body. *B* Oligodendrocyte near an apical dendritic shaft (den) of a cortical pyramidal neuron.

Early in development, glial cells assume a radial distribution in the de-veloping neural tube, with their processes extending from the central canal to the pial surface (for more detail on the role of glia in development, see Chapter 5). At this stage, the glia are termed *radial glia*. Migrating neurons follow these radial glial processes out from the proliferative zone into the area where they will differentiate. After migration is complete, these glial cells retract their long radial processes and differentiate into astrocytes.

In addition to the major glial types, there are specialized types of glial cells that are unique to particular brain regions. All of these are thought to be derivatives of the astrocyte lineage.

Golgi epithelial cells or *Bergmann glia* are found in the cerebellar cortex; their cell bodies and nuclei are positioned just below the Purkinje cell layer, and they send a long process radially through the molecular layer to the pial surface (Fig. 2.9). These processes are termed *Bergmann fibers*. During early development, migrating granule cells follow the processes of Bergmann glia from the external granule layer to the internal granule layer.

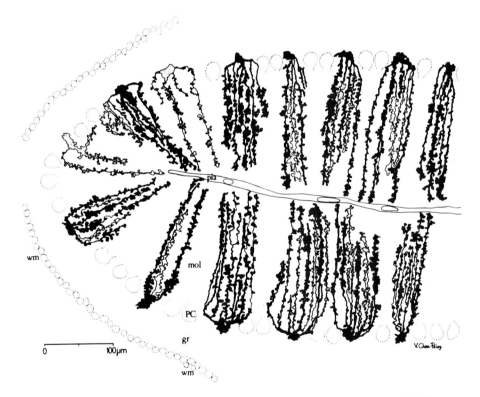

FIGURE 2.9. Golgi epithelial cells or Bergmann glia in the cerebellum. This illustration shows a composite drawing of 14 Golgi epithelial cells that were stained by the rapid Golgi method. Each cell sends long processes (Bergmann fibers) to the pial surface (pia). mol = molecular layer, PC = Purkinje cell layer; gr = granule layer, wm = white matter. [From Palay SL, Chan-Palay V (1974) *Cerebellar Cortex Cytology and Organization*. Berlin, Springer-Verlag.]

Müller radial cells of the retina or *retinal gliocytes* also have features in common with astrocytes. Like the Bergmann glia, these are large cells; their cell bodies are found in the internal nuclear layer near the cell bodies of the ganglion cells. These cells send out elaborate processes oriented radially to the retinal layers. These processes fan out to form the *external limiting lamina* of the retina, between the lamina of rods and cones and the external nuclear lamina containing the nuclei of rod and cone cells. The processes of Müller cells also encapsulate photoreceptor processes, neural elements, and extend to form the internal limiting membrane on the vitreal aspect of the retina.

Pituicytes of the neurohypophysis are also thought to be specialized astrocytes. The processes end mostly on endothelial cells of the vascular sinuses.

To some extent, the morphological distinctions between different types of glial cells become blurred in some situations. For example, after injury, astrocytes undergo a morphological transformation as they begin to engage in phagocytosis of degeneration debris (see below). One indication of this transformation is a dramatic increase in glial filaments. Thus, ''protoplasmic astrocytes'' that become reactive in response to injury are morphologically very similar to fibrous astrocytes. Nevertheless, the differences between fibrous and protoplasmic astrocytes may extend beyond their morphology. For example, cultures of astrocytes prepared from white matter yield two kinds of GFAP-positive cells (one of which is process-bearing, and resembles fibrous astrocytes in vivo). Cultures of astrocytes prepared from immature cerebral cortex yield a different type of GFAP-positive cell that exhibits a polygonal morphology; these cells are thought to represent the protoplasmic astrocyte lineage. The results suggest that the differences between glial types may actually reflect fundamental differences in cell type.

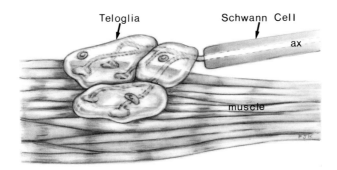

FIGURE 2.10. Schwann cells at the terminals of motor axons. Schwann cells surrounding the neuromuscular junction are termed teloglia. ax = axon.

Glial Cells of the Peripheral Nervous System (PNS)

Schwann cells are analogous to oligodendrocytes and are found in all types of peripheral nerve. In myelinated nerves, the Schwann cells form the myelin sheath. Unmyelinated nerves have an envelope of Schwann cell membrane. Schwann cells also encapsulate the terminal arborizations of motor nerves. In this location, they are also called *teloglia* (Fig. 2.10).

Functional Roles of Glial Cells

A number of functional roles have been proposed for glial cells; some of these roles are well established, and others are based upon reasoned speculation.

Functions of Oligodendrocytes and Schwann Cells

Schwann cells and oligodendrocytes are thought to contribute to the metabolic support of the axons that they envelop; they also produce the myelin that speeds conduction. The myelin is actually composed of the plasma membrane of the glial cell, which is wrapped several times around the axon forming the characteristic myelin lamellae. In general, Schwann cells of the PNS form myelin around only one axon, and the cell body of the Schwann cell lies adjacent to the axon that it myelinates. In unmyelinated nerves, many individual axons may be enveloped by the cytoplasm of a single Schwann cell (Fig. 2.11). In the CNS oligodendrocytes send out processes that contact several axons, sometimes at a distance from the cell body of the oligodendrocyte; the wrappings of the myelin sheath are formed from the plasma membranes of these oligodendrocyte processes (Fig. 2.12).

The myelin sheath acts as insulation around the axon; ionic currents associated with the nerve impulse cannot flow across the myelin and therefore flow across the nerve membrane at *Nodes of Ranvier*. It is this property that results in *saltatory conduction,* where the action potential jumps from one node to the next, thus increasing the net speed of conduction.

The Regulation of Myelin Formation

The extent of myelination varies considerably in different systems. For example, in the PNS, motor and sensory nerves are heavily myelinated, whereas sympathetic nerves are unmyelinated. There are also differences in the extent of myelination of different pathways in the CNS. Since demyelinating disease is fairly common and severely debilitating, it is important to understand how the extent of myelination is regulated. The question is whether it is the myelinating cell or the axon that determines the extent of myelination. In other words, are there different types of glial

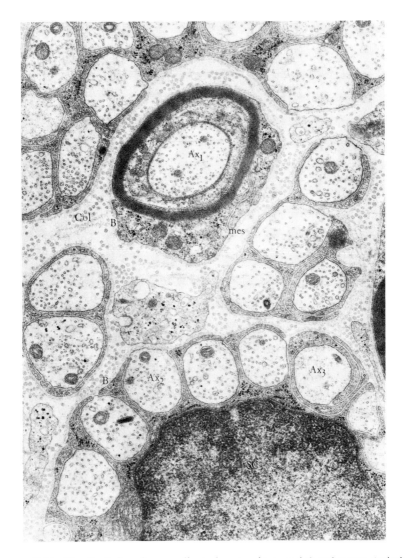

FIGURE 2.11. Myelinated and unmyelinated axons in a peripheral nerve (sciatic nerve of adult rat). Ax_1 is a myelinated axon sectioned transversely through the paranodal region. The paranodal region can be identified because of the close apposition between the plasma membranes of the axon and the sheath and thick layer of Schwann cell cytoplasm between the axon and myelin lamellae. Termination of the myelin lamellae at the mesaxon (mes) is also indicated., The other axons in the field are unmyelinated, but nevertheless are surrounded by fingers of cytoplasm from Schwann cells. This relationship is particularly well illustrated by the Schwann cell in the lower portion of the figure (SC), which encloses several axons. Ax_3 is surrounded completely by a process from another Schwann cell. Between the Schwann cells is the collagen (Col) of the endoneurium. [From Peters A, Palay SL, Webster H deF (1976) *The Fine Structure of the Nervous System: The Neurons and Supporting Cells.* Philadelphia, WB Saunders.]

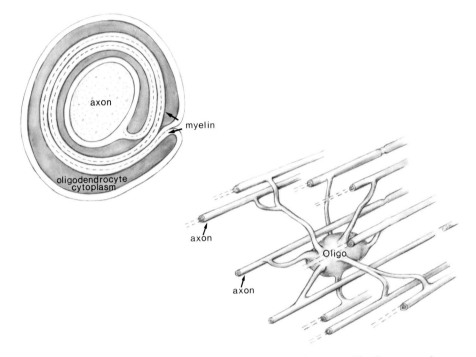

FIGURE 2.12. Formation of CNS myelin by oligodendrocytes. The lower portion of the figure illustrates how a single intrafascicular oligodendrocyte gives rise to processes that form the myelin sheaths around several axons. The upper portion of the figure illustrates a relatively early stage in the process of myelin formation to emphasize the fact that the myelin actually consists of several wrappings of the plasma membrane of the oligodendrocyte. As the myelin matures, the membrane wraps become more closely apposed, eliminating the fingers of oligodendrocyte cytoplasm.

cells in heavily myelinated nerves (in contrast to lightly myelinated) or is the capability of the glial cell the same, with the difference being in the axons? Recent work by A. Aguayo and his colleagues has provided some answers to this question (Aguayo et al., 1979).

Aguayo and his colleagues used heterologous transplants to study the formation of myelin by Schwann cells. They removed a segment of the sciatic nerve, which is usually heavily myelinated, and replaced it with a segment from a sympathetic nerve, which is lightly myelinated. In the graft the axons degenerate, since they have been separated from their cell bodies; the Schwann cells survive, however, and can remyelinate axons that regenerate into the graft. When sciatic nerve axons regenerated into a graft containing Schwann cells from sympathetic nerves, a heavy myelin sheath was formed. Thus, Schwann cells from a sympathetic nerve that do not normally produce heavy myelin sheaths can be induced to do so

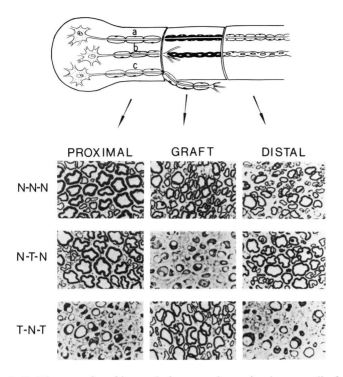

FIGURE 2.13. The use of grafting techniques to determine how myelin formation is regulated. The experimental strategy is illustrated in the upper portion of the figure. A section of the nerve is removed and replaced with a graft from either the same animal or a different animal. In the early stages the axons distal to the first cut degenerate, leaving denervated Schwann cells. Axons regenerating from the proximal stump are ensheathed by Schwann cells from the graft within the graft proper and by Schwann cells of the host in the distal segment. Any axons that grow outside the perineurium are ensheathed by Schwann cells that migrate from the proximal stump. The lower portion of the figure illustrates the use of the grafting techniques to evelute the site of the defect in the *trembler* mutation of mice. When segments of normal nerves are grafted into nerves from normal mice, (N-N-N), normal myelin is formed in both host and graft. When segments of nerves from trembler mice are grafted into nerves from normal mice (N-T-N), the myelin within proximal and distal segments is normal, but the fibers in the graft lack myelin. When segments of normal nerves are grafted into nerves from trembler mice (T-N-T), normal myelin is formed within the graft, even though the myelin of the host is abnormal. [After Aguayo AJ, Bray GM, Perkins CS (1979) Axon-Schwann cell relationships in neuropathies of mutant mice. *Ann NY Acad Sci* 317: 512–533.]

if they come into contact with axons of the appropriate type. This indicates that the extent of myelin formation is regulated by the axon.

The technique of grafting has provided an important tool for exploring the sites of defects in genetic diseases that result in abnormalities in myelin formation. For example, Aguayo was able to show that the defect in the "trembler" mutant is in the Schwann cells rather than in the axons (Fig. 2.13). When segments of nerves from trembler mice were grafted into normal nerves, axons that regenerated through the grafts exhibited abnormal myelination. Similarly, when segments of nerves from normal mice were grafted into trembler nerves, the axons that regenerated into the graft exhibited normal myelin. Using similar techniques, but transplanting human nerve segments into immunosuppressed mice, it was found that the human genetic disorder *metachromatic leucodystrophy* was also due to a defect in Schwann cells. These kinds of transplantation techniques provide a potentially important tool for investigating the sites of defects in other disorders of myelination.

Role of Schwann Cells in Regeneration and Repair

Schwann cells appear to facilitate axonal regeneration and guide regenerating axons. Santiago Ramon y Cajal, in his classic studies of degeneration and regeneration in the nervous system, observed that axonal regeneration in all parts of the CNS is quite limited, whereas in the PNS, axon growth readily occurs (Cajal, 1928). Since a major difference between nerves in the PNS and CNS is the presence of Schwann cells in the PNS, Cajal proposed that Schwann cells stimulate nerve regeneration.

This hypothesis has been confirmed by recent experiments. For example, using the grafting techniques described above, Albert Aguayo and his colleagues found that when CNS nerves containing only oligodendrocytes (the optic nerve) are grafted into the sciatic nerve, regeneration of the axons of the sciatic nerve is very limited. Grafts that contain Schwann cells support regeneration, however. Similarly, when peripheral nerves containing Schwann cells such as the sciatic nerve are implanted into the CNS, which normally exhibits little or no regeneration, a vigorous regenerative response is observed (see Chapter 8). These grafts offer considerable hope for the eventual repair of CNS injuries, particularly injuries to the spinal cord. It may eventually be possible to use grafts of peripheral nerve to bridge areas of damage and thus permit the reconnection of centers that are disconnected as a consequence of trauma.

Perineuronal Oligodendrocytes

The close relationship between perineuronal oligodendrocytes and neurons suggests that the oligodendrocytes interact with the neurons. Indeed, the term "satellite cell" seems particularly appropriate for the perineuronal oligodendrocyte. Although the close physical relationship between perineuronal oligodendrocytes and neurons certainly suggests an interdependence, the nature of any interaction between these two cell types is currently unknown.

Functions of Astrocytes

Astrocytes are thought to play an important role in (1) providing a mechanism for the exchange of materials between capillaries and neurons, (2) providing nutritive or trophic support for neurons, (3) providing a buffering mechanism by regulating extracellular ionic composition and by taking up neurotransmitter substances that have been released by neurons, (4) removing degeneration debris after injury, particularly degenerating synaptic terminals, and (5) contributing to cellular compartmentation in densely packed neuropil zones by separating neuronal processes. In addition, during development, astrocytes appear to play an important role in guiding migrating neurons and perhaps in providing a substrate for long-distance growth of axons (see Chapter 5).

Exchange of Substances Between Capillaries and Neurons

The close apposition between neurons and glia and the presence of glial end feet on capillaries have led to the suggestion that glia aid in the distribution of nutritive substances to neurons. This hypothesis, which was elaborated by Camilio Golgi in about 1883, has remained quite popular. In forming end feet on the capillaries, the glia seem ideally positioned to receive substances from the capillaries; indeed, because glial end feet completely surround capillaries, it seems likely that most substances entering the CNS pass through the cytoplasm of the astrocyte. Since these same astrocytes also contact neurons, the astrocytes are ideally positioned to exchange materials. Although this is an extremely attractive and popular hypothesis, there is little direct experimental evidence for the hypothesis, and it thus remains speculative.

   In addition to conveying materials from capillaries to neurons, astrocytes might also convey material in the opposite direction from the neurons to the capillaries. These materials could be by-products of cellular metabolism or could represent a means that would allow neurons to communicate their metabolic needs to the vasculature. An interesting and important aspect of cerebral blood flow is that it is closely coupled to neuronal metabolism, which is, in turn, closely coupled to neuronal activity. It is not currently known how this coupling is accomplished, but from the structural point of view, astrocytes seem ideally positioned to mediate such communication.

Nutritive or Trophic Role of Astrocytes

The most direct evidence for a nutritive or trophic role for astrocytes comes from studies of neurons grown in tissue culture. When neurons are removed from very young animals and grown in explant cultures, where both neurons and glia survive, the neurons survive and differentiate in a fashion that is quite comparable to the in vivo situation. However, when the neurons are plated at low density in the absence of glial cells, their survival is generally limited. The survival of some neurons can be greatly enhanced by co-culturing the neurons with glia, suggesting that the glia provide some material that contributes to neuronal survival (trophic substances).

In evaluating the potential trophic role of glial cells, it is important to distinguish between factors that are produced by glia and those that are produced by other neurons or by target cells upon which neurons terminate. For example, there is now excellent evidence for target-derived factors. The best known of these is nerve growth factor (NGF), which is produced in tissues that receive sympathetic innervation. Sympathetic and sensory neurons depend upon NGF for their survival; in the absence of NGF, these neurons die (for further discussion, see Chapter 5). Other target-derived factors may also exist, and evaluating the potential role of glia in providing trophic support requires that the source of any trophic factor be identified. In turn, this identification depends upon the existence of appropriate assays.

In the case of NGF, sensitive assays are available. Antibodies have been available to the molecule for some time; and the gene for NGF has been cloned, permitting an analysis of the sites of synthesis. However, the potential trophic molecules that might be released by astrocytes are unknown; thus, one must use indirect means to assay them. Considerable progress toward assaying and then identifying unknown factors has been made by a number of investigators. For example, S. Varon and his colleagues have developed tissue culture bioassay systems by taking advantage of the fact that many types of neurons cannot survive in culture in the absence of exogenous factors (Varon et al. 1983). For example, some neurons depend upon extracts from CNS tissue for their survival. When these cultures are grown in the presence of different dilutions of tissue extract, a "dose-response" curve can be constructed. At high concentrations of a given factor, the survival of the neurons in the test culture is maximal; as the factor is diluted, the number of surviving neurons declines in a dose-dependent fashion until some background value is reached. The dose-response curve then allows for a determination of the amount of the "factor" in different tissues.

Using such bioassays, Varon and his colleagues obtained evidence for a number of different factors that have different biological effects. Some factors promote neuronal survival; these are termed *neuronotrophic factors*. Other factors promote the outgrowth of neuronal processes *(neurite promoting factors)*. There is also evidence for *inhibitory factors* that appear to interfere with the survival of neurons. Different types of neurons appear to have very different sensitivities to the added factors, suggesting a high degree of specificity of action. Because some of the factors are present in extracts and conditioned media from glial cells, at least some neuro-trophic factors appear to be of glial origin.

There is also evidence that diffusable trophic substances are released in vivo. For example, factors of some sort accumulate in wound cavities in the brain. These factors can be harvested by placing gel foam in the cavity, and this material substantially enhances the survival of a variety of neuron types in culture. It is presumed that these soluble factors arise from glial cells, although a neuronal origin for some or all of the factors cannot be excluded. Further progress in this important field will almost certainly require the identification of the factors and the production of appropriate molecular probes (antibodies) that will permit an analysis of the site of synthesis and release of the factors. This important work is very difficult because of the very low concentrations of the growth factors.

Role of Astrocytes in Buffering the Extracellular Environment of the Brain

The proper function of neurons depends upon the ionic composition of the extracellular fluid. In particular, high concentrations of potassium ions depolarize neurons and can induce abnormal excitability. Since neuronal activity leads to potassium efflux from neurons, which leads to an ac-cumulation of extracellular potassium, there must be a mechanism for the removal of this excess potassium. Indeed, this mechanism must be a highly effective one, since it has been estimated that potassium efflux associated with neuronal activity can raise extracellular potassium to more than 50 times its normal level in the absence of special clearance mechanisms.

There is good evidence that astrocytes provide the principal means for removing excess potassium from the extracellular medium. Astrocytes appear to accomplish this task by accumulating the potassium that is re-leased into the extracellular medium through *passive spatial buffering* and as a result of *active uptake mechanisms*. The accumulation of potassium by astrocytes has been demonstrated in a number of ways, most con-vincingly through direct measurements of increases in intracellular K^+ concentrations using ion-sensitive microelectrodes.

Passive spatial buffering was initially proposed as a result of studies of the physiological properties of glial cells. Taking advantage of the very large and identifiable glial cells in the optic nerve of the mudpuppy and the ganglia of the leech, Orkand et al. (1960) found that astrocytes are highly permeable to potassium; indeed, their permeability to potassium is considerably higher than to any other ion. As a result, the membrane potential of astrocytes depends almost entirely upon potassium concen-tration (see below). Because of their high permeability to potassium, K^+ readily enters glia at focal areas of high extracellular concentration and leaves the glial cells at areas of normal concentration by way of a current

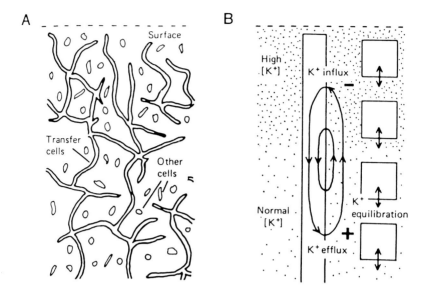

FIGURE 2.14. Passive spatial buffering by astrocytes. Passive spatial buffering could occur in any area where there existed a network of electrically coupled cells (transfer cells). *A* Illustrates the arrangement in the cerebral cortex, where astrocytes would represent the transfer cells by virtue of their tight junctions with one another. *B* Because of the coupling via tight junctions, transfer cells can be represented as an extended tube with low intracellular resistance. Other cells that are not coupled (neurons) are represented by the boxes. *Arrows* indicate the pattern of current and ion fluxes that would occur in response to focal increases in extracellular potassium concentration such as might arise following local neuronal activity. [From Gardner-Medwin AR (1983) Analysis of potassium dynamics in mammalian brain tissue. *J Physiol* 335: 393–426.]

loop, with intracellular and extracellular ionic current flow completing the circuit (Fig. 2.14). This mechanism is made even more effective because glial cells are electrically coupled to one another through low resistance gap junctions. Thus, ionic current that flows into one glial cell is readily dispersed through these low-resistance junctions to other glial cells.

In addition to the passive spatial buffering, some of the accumulation of excess potassium appears to be due to active uptake. This conclusion is based upon the fact that the accumulation is partially blocked when the sodium-potassium pump is disrupted by ouabain.

Astrocytes also possess uptake mechanisms for neurotransmitters and may participate in removing neurotransmitters from the extracellular space after they have been released by neurons. The relative contributions of glial cells versus neuronal reuptake mechanisms (see Chapter 3) in physiological settings may vary in different situations and with different types of neurotransmitter.

Role of Astrocytes in Removing Degeneration Debris After Injury

After damage to the nervous system, which results in the degeneration of neural tissue, astrocytes play an important role in removing degenerating debris (*phagocytosis*). Unlike neurons, at least some types of astrocytes retain the capacity to divide throughout the life of the organism; after damage, they divide, undergo considerable hypertrophy, and occupy the spaces left vacant by dying tissue, thus forming *glial scars*. In contrast to microglia, which are thought to migrate to sites of damage from nearby vascular elements (see below), astrocytes are thought to divide in place and then undergo hypertrophy.

When astrocytes undergo hypertrophy in response to injury, they are termed *reactive astrocytes*. In the reactive phase, astrocytes enlarge considerably, increase the length and extent of their processes, and substantially increase their production of glial filaments composed of GFAP (Fig. 2.7). When protoplasmic astrocytes become reactive, their glial filaments increase substantially, and the protoplasmic astrocytes thus assume some of the characteristic fibrous astrocytes. In areas where neurons or their processes are degenerating, astrocytes can be found engulfed with degeneration debris. Astrocytes seem to play an especially important role in the phagocytosis of degenerating axons and synapses. They engulf the degenerating synapse, clearing the postsynaptic site of the dying synaptic terminal (Fig. 2.15).

The removal of degenerating terminals by astrocytes may play an important role in nervous system repair after injury. Although regeneration does not often occur in the CNS, other types of growth do take place. For example, neurons that lost their synaptic connections because of a distant lesion are often reinnervated as a result of *sprouting* of nearby axons that are undamaged (see Chapter 8). Before neurons can be reinnervated, the dying synapses must be removed; this is accomplished by reactive astrocytes. There is some evidence that the phagocytosis of degenerating synapses is, in fact, "rate limiting" for reinnervation; when glial function is impaired, such as in aged brains, the rate of removal of degenerating synapses is impaired, and the extent of reinnervation of denervated cells is also reduced.

The response of glial cells to injury can be useful for diagnosing sites of degeneration following injury. Since glial cells accumulate around dying neurons or their processes, collections of glial cells and glial scars can reveal microscopic injuries to the nervous sytem that would otherwise be invisible.

Cellular Compartmentation

The fact that glial cells encapsulate neuronal processes also suggests that the glia may play a role in "insulating" nerve processes from one another. Very early on, Santiago Ramon y Cajal proposed that glia prevented cross-

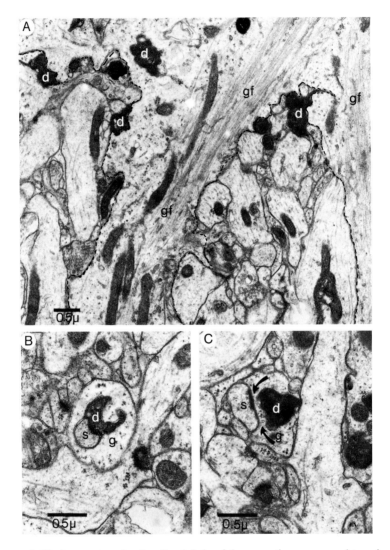

FIGURE 2.15. Astrocytes clearing the debris of degenerating axons and terminals. *A* Cytoplasm of an astrocyte in an area of neuropil that contains many degenerating terminals as a result of a lesion that destroyed the afferents to the zone. Degeneration debris (d) can be seen in the neuropil and within the cytoplasm of the astroycte. Note the extensive glial filaments in this reactive astrocyte (gf). *B* A glial process (g) surrounds a degenerating synaptic terminal that is still in contact with a spine. The spine is probably still connected to its parent dendrite via a spine neck that is not visible in this section. *C* A glial process (g) surrounds a degenerating synaptic terminal (d) that is immediately adjacent to an empty spine (s). *Arrows* indicate the hypothesized action of the glial process as it removes the degenerating terminal from the postsynaptic site.

talk by preventing current flow between adjacent neural elements. On the basis of our present understanding of current flow, it now seems unlikely that direct electrical interactions would occur between neuronal processes, unless there is a large area of apposition. However, glial cells could help to prevent the diffusion of various substances (ions and neurotransmitters) from one neuronal site to another. This may be especially important in densely packed neuropil zones, where synaptic terminals lie adjacent to one another.

The Role of Glia in Neuronal Migration and Development

Studies of the development of the cerebral cortex, hippocampus, and particularly the cerebellar cortex, have suggested that glial cells play an important role in neuronal migration. In early development these areas contain *radial glia,* which have long processes that extend for the entire thickness of the neural tube from the central canal to the surface (see Chapter 5). Neurons follow the processes of these radial glia as they migrate from the proliferative matrix out into the substance of the CNS. In the cerebellar cortex, a similar process occurs in the migration of the granule cells. These cells are produced in the external granule layer and then migrate down through the molecular layer past the Purkinje cell layer to the granule layer. In this migration, the granule neurons follow the long radial process of Bergmann glia. Thus, the glial cells are thought to provide a guiding matrix around which subsequent neuronal organization takes place.

Radial glial cells may also guide the initial growth of long axonal tracts. Marcus Singer and his associates at Case Western Reserve University have proposed a ''substrate hypothesis'' for axon growth in long tracts (i.e., the spinal cord and optic nerve Singer et al. 1979). Early in development, the processes of radial glia form potential ''channels'' along the length of the optic nerve and spinal cord. During the initial growth of long tracts, axons seem to grow along these radial glial processes.

Electrophysiological Properties of Astrocytes

Physiological studies of identified astrocytes in the optic nerve of the mudpuppy and in ganglia of the leech have revealed some of the electrophysiological properties of glia (for a comprehensive review, see Orkand 1977). It has been found that these astrocytes have higher resting membrane potentials than neurons and have electrically inexcitable membranes. Indeed, the glial membrane behaves passively even when displaced over a range of 200 mV. Most importantly, the membrane potential of an astrocyte varies as a direct function of potassium ion concentration in the extracellular medium (Fig. 2.16). Changes in ions other than K^+ have a negligible effect on the membrane potential. Because the membrane potential is determined entirely by the potassium ion concentration in the extracellular medium, glia can be said to operate as potassium electrodes.

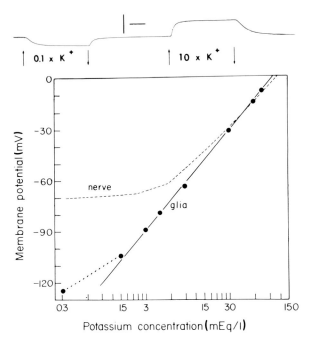

FIGURE 2.16. Electrophysiology of glia; dependence of glial membrane potential on K^+ ion concentration. Astrocytes in the optic nerve of *Necturus* were impaled by micropipette electrodes, and the membrane potential was determined at a variety of K^+ concentrations. The top of the figure illustrates the changes in the membrane potential of the astrocyte following changes in K^+ concentration in the bath. The lower portion of the figure plots the relationship between glial membrane potential and extracellular potassium concentration. Note that the membrane potential of the astrocyte varies directly with the extracellular potassium ion concentration. The dotted line indicates a similar plot for frog myelinated nerve fibers. [*Top* From Kuffler SW, Nicholls JG, Orkand RK (1966) Physiological properties of glial cells in the central nervous system of amphibia. *J Neurophysiol* 29: 768–787. *Lower portion* Orkand RK (1971) Neuron-glia relations and control of electrolytes, in Siesjo BK, Sorenson SC (eds): *Ion Homeostasis of the Brain* (ABSY 3), © 1971 Munksgaard International Publishers Ltd., Copenhagen, Denmark.]

Intracellular recordings from astrocytes in the optic nerve of the mud-puppy reveal that there are slow changes in the glial membrane potential when surrounding axons are activated (Fig. 2.17). These changes could not be due to direct current flow between neurons and glia because the changes in the membrane potential of the glia occur over a different time scale than those of the electrical events in the neurons. The changes are, however, consistent with the interpretation that the glial cells are depolarized as a result of the efflux of K^+ from the active axons. Direct evidence for this interpretation has been obtained by studying the activity-induced

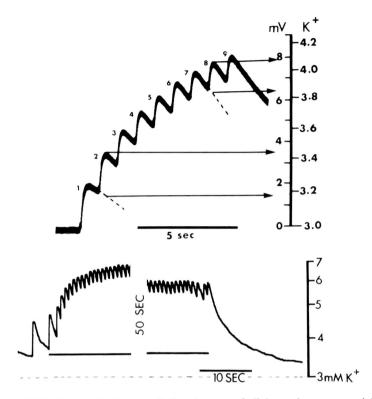

FIGURE 2.17. Electrophysiology of glia; changes of glial membrane potential consequent to neural activity. *Top* Intracellular recordings were made from glial cells in the optic nerve of *Necturus* during activation of optic nerve at 1/s. Note successive depolarization of glial cell. The values of external K$^+$ that would produce equivalent depolarization (calculated from the Nernst equation) are shown on the right. [From Orkand RK, Nicholls JG, Kuffler SW (1966) The effect of nerve impulses on the membrane potential of glial cells in the central nervous system of amphibia. *J Neurophysiol* 29, 788–806.] *Bottom* Changes in extracellular potassium concentration in cat cortex with electrical stimulation as revealed by potassium sensitive microelectrodes positioned in the extracellular space. [From Prince DA, Lux HD, Neher E (1973) Measurement of extracellular potassium activity in cat cortex. *Brain Res* 50: 489–495.]

changes in glial membrane potential when extracellular K$^+$ is varied systematically. If the slow changes in glial membrane potential are due to an efflux of K$^+$ from neurons, then the changes in glial membrane potential should be *reduced* when resting extracellular K$^+$ is elevated and *enhanced* when resting extracellular K$^+$ is decreased. This is true because the K$^+$ that is released from active neurons should add logarithmically to the K$^+$ already present extracellularly. It was found that the slow potentials did vary as predicted by the hypothesis that the depolarization that occurred in the glial cells was a direct result of the increase in extracellular K^+ consequent to neuronal activity. Thus, it is thought that the membrane potential of glial cells fluctuates in response to the activity of nearby nerve cells as a result of changes in extracellular K^+. It is not clear whether this sensitivity to extracellular K^+ plays a role in the functional activities of glia, but it certainly represents an attractive mechanism through which glial function could be regulated by neuronal activity.

Neurotransmitter Receptors on Astrocytes

Recent studies have revealed that astrocytes possess receptors for neurotransmitters, neuromodulators, and hormones. Presumed astrocytes in culture (that is GFAP-positive cells) express a wide variety of neurotransmitter receptors, as revealed by binding studies using specific ligands. In addition, physiological and biochemical studies have revealed that activation of these receptors on astrocytes has similar effects as neurotrans-

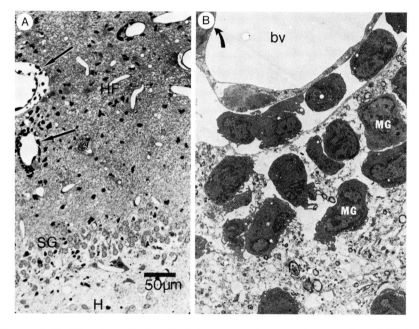

FIGURE 2.18. Migration of microglia into areas undergoing degeneration. *A* Light micrograph of a 1μm section through the dentate gyrus of the rat after injections of colchicine that led to the degeneration of neurons. Microglia can be seen surrounding blood vessels (*arrows*). *B* Electron micrographs of microglia (MG) surrounding blood vessels in the dentate gyrus. [Reprinted with permission from Goldschmidt RB, Steward O (1982) Neurotoxic effects of colchicine: differential susceptibility of CNS neuronal populations. *Neuroscience* 7: 695–714, Copyright 1982, Pergamon Press plc.]

mitters operating on neurons, i.e., the receptors can lead to alterations in ion flux or generation of second messengers such as cyclic adenosine monophosphate (cAMP). An important caveat is that most of these studies involve glial cells grown in culture, which may express different characteristics than glial cells in situ. Nevertheless, enough substantiating evidence exists to suggest that the conclusions obtained from studies of cells in culture are correct in principle and that glial cells do possess receptors for many neurotransmitter substances. The role of these receptors and second messenger systems in modulating glial function is not known, but it is certainly plausible that these receptors play an important role in mediating neuronal-glial interactions.

Functions of Microglia

Microglia are traditionally thought to be quiescent in normal brain, serving as potential phagocytes that can respond when needed to remove damaged tissue after injury. As noted above, microglia have ultrastructural characteristics that are similar to those of macrophages and are capable of proliferating in response to injury. Furthermore, unlike other glial cells, the microglia are capable of migrating within the CNS and accumulate in areas undergoing degenerative changes as a consequence of injury. Early in the course of degeneration, microglia can be seen surrounding blood vessels in a configuration that strongly suggests that they originate from the vasculature and migrate into regions containing damaged cells (Fig. 2.18). It is not clear whether microglia have other roles in the nondamaged nervous system.

Gene Expression in Glial Cells

While the three principle types of CNS glia clearly represent well-defined cell types, there are structural and biochemical distinctions between subgroups within each class. This is particularly true of the astrocytes. As noted above, astrocytes exhibit quite different morphologies, depending upon their location. Even within a single structure, there are structural and biochemical differences between subtypes. For example, in the optic nerve, the astrocytes that form the glial limiting membrane possess different membrane glycoproteins than the fibrous astrocytes in the core of the nerve.

The techniques of cellular and molecular biology that have been so important for advancing our knowledge about neurons have also provided information about glial cells. Several *glial-specific proteins* have been identified, some of which have clearly defined functional roles. Thus, glial fibrillary acidic protein (GFAP) is the protein that comprises glial filaments; the protein is specific to astrocytes, and thus serves as a molecular marker for these cells. *Myelin basic protein* (MBP) and *myelin proteolipid protein* (PLP) are major protein constituents of CNS myelin, and thus serve as molecular markers for oligodendrocytes.

The genes for several of the glial proteins have been cloned, and their mode of expression has been evaluated. It is known for example that there are several distinct mRNA's for MBP and PLP that are produced by alternative RNA splicing. There are well-characterized genetic defects that involve the MBP gene (the *shiverer* mutation in mice), and the PLP gene (the *jimpy* mutation in mice). An extremely interesting feature about the expression of MBP is that it may involve a targeted transport of the MBP mRNA to the distal processes of the oligodendrocyte, where the MBP protein appears to be synthesized within portions of the cytoplasm very near its site of insertion into the myelin membrane.

Regulation of Glial Gene Expression by Neuron-Glial Interations

Not surprisingly, some aspects of glial gene expression are regulated through neuronal-glial interactions. The mechanisms for the regulation are not always understood, but the effects are clear cut. For example, as noted above, axons determine the extent of myelin formation by Schwann cells; this must involve a regulation of the expression of myelin proteins. Similarly, when astrocytes become reactive in response to injury, the glial hypertrophy unquestionably requires alterations in gene expression, including a considerable increase in the production of GFAP. Finally, glial enzyme expression also depends upon neuron-glial interactions. For example, Bergmann glia normally express very high levels of glycerol-3-phosphate dehydrogenase (GPDH). However, in mutant cerebella lacking Purkinje cells, GPDH expression by Bergmann glia is greatly reduced, although the Bergmann glia themselves are intact. The molecular mechanisms for these interactions are not known, but the effects are clear cut. Given the high level of interest in these issues, there is every reason to anticipate major advances in our understanding of the mechanisms of glial gene expression in the near future.

Conclusion

The experimental efforts over the last 75 years have revealed that glial cells are not simply the glue that holds the nervous system together. Glial cells clearly interact with neurons in many ways, with each type of glial cell playing a particular role. Some of the roles of glial cells are becoming clear, particularly the role of Schwann cells and interfascicular oligodendrocytes in myelin formation and the role of astrocytes in regulating the brain microenvironment and in phagocytosing degeneration debris after injury. Much is still to be learned, however, and the functional role of some types of glia (i.e., perineuronal oligodendrocytes) is still largely enigmatic.

Supplemental Reading

General

Federoff S, Vernadakis A (eds) (1986) *Astrocytes; Development, Morphology, and Regional Specialization of Astrocytes, vol 1.* New York, Academic Press

Federoff S, Vernadakis A (eds) (1986) *Astrocytes; Development, Biochemistry, Physiology, and Pharmacology of Astrocytes, vol 2. New York, Academic Press*

Grisar T, Hertz FG, Norton WT et al. (1987) *Dynamic Properties of Glial Cells II: Cellular and Molecular Aspects.* Oxford, England, Pergamon Press

Jacobson M (1978) *Developmental Neurobiology.* New York, Plenum Press

Kuffler SW, Nicholls JG (1976) *From Neuron to Brain.* Sunderland MA, Sinauer Assoc Inc

Purves D, Lichtman JW (1985) *Principles of Neural Development.* Sunderland MA, Sinauer Assoc Inc

The Cytology of Glia

Raff MC, Abney ER, Cohen J et al. (1983) Two types of astrocytes in culture of developing rat white matter: differences in morphology, surface gangliosides, and growth characteristics. *J Neurosci* 3:1289–1300

Ramon y Cajal S (1911) *Histologie du Système Nerveux de l'Homme et des Vertébrés* (2 vols) (Axoulay L, trans) Reprinted by Instituto Ramon y Cajal del CSIC, Madrid, 1952–1955

Peters A, Palay SL, Webster DeF (1976) *The Fine Structure of the Nervous System: The Neurons and Supporting Cells. Philadelphia, WB Saunders Co*

Myelin Formation

Aguayo AJ, Bray GM, Perkins CS, et al. (1979) Axon-sheath cell interactions in peripheral and central nervous system transplants. *Soc Neurosci Symp* 4:361–383

Spencer PS (1979) Neuronal regulation of myelinating cell function. *Soc Neurosci Symp* 4:275–321

Glial Cells in Development, Degeneration, and Regeneration

Keynes RJ (1987) Schwann cells during development and regeneration: leaders or followers? *Trends Neurosci* 10:137–139

Aguayo A, David S, Richardson P, et al (1982a) Axonal elongation in peripheral and central nervous system transplants. *Adv Cell Neurobiol* 3:215–234

Aguayo AJ, Richardson PM, Benrey M (1982b) Transplantation of neurons and sheath cells-a tool for the study of regeneration, in Nicholls JG (ed): *Repair and Regeneration of the Nervous System. Life Sciences Research Report 24.* Berlin, Springer-Verlag, pp 91–106

Ramon y Cajal S (1926) Degeneration and regeneration of the nervous system, vols I and II. Republished by Hafner Publishing Co., London, 1968

Singer M, Nordlander RH, Egar M (1979) Axonal guidance during embryogenesis and regeneration in the spinal cord of the newt: the blueprint hypothesis of neuronal pathway patterning. *J Comp Neurol* 185:1–21

Gall C, Rose G, Lynch G (1979) Proliferative and migratory activity of glial cells in the partially deafferented hippocampus. *J Comp Neurol* 183:539–550

Astrocytes and Neurotrophic Factors

Banker GA (1980) Trophic interaction between astroglia and hippocampal neurons in culture. *Science* 209:809–810

Lindsay RM, Barber PC, Sherwood MRC et al (1982) Astrocyte cultures from adult rat brain. Derivation, characterization and neurotrophic properties of pure astroglial cells from corpus callosum. *Brain Res* 243:329–343

Nieto-Sampedro M, Lewis ER, Cotman CW et al. (1982) Brain injury causes a time-dependent increase in neuronotrophic activity at the lesion site. *Science* 217:860–861

Perez-Polo JR, deVellis J, Haber B (eds) (1983) Growth and Trophic Factors. *Prog Clin Biol Res* 118

Thoenen H, Korsching S, Barde YA et al. (1983) Quantitation and purification of neurotrophic molecules. *Cold Spring Harbor Symp Quant Biol* 48:679–683

Varon S Adler R, Manthorpe M et al. (1983) Culture strategies for tropic and other factors directed to neurons, in Pfeiffer SE (ed): *Neuroscience Approached through Cell Culture,* vol 2. Boca Raton, CRC Press, pp 53–77

Neurotransmitter Receptors on Astrocytes

Murphy S, Pearce B (1987) Functional receptors for neurotransmitters on astroglial cells. *Neuroscience* 22:381–394

Electrophysiological Properties of Astrocytes

Ransom BR, Carlini WG (1986) Electrophysiological properties of astrocytes, in Federoff S, Vernadakis A (eds): *Astrocytes; Development, Biochemistry, Physiology, and Pharmacology of Astrocytes* vol 2. New York, Academic Press, pp 1–49

Orkand RK (1977) Glial cells, in Brookhart JM, Mountcastle VB (section eds). *Handbook of Physiology, Section 1: The Nervous System.* Baltimore, Williams and Wilkins, pp 855–875

Orkand RK, Nicholls JG, Kuffler SW (1966) The effect of nerve impulses on the membrane potential of glial cells in the central nervous system of amphibia. *J Neurophysiol* 29:788–806

CHAPTER 3

Interneuronal Communication I: Neurotransmitters

Introduction

Neurons communicate with each other and with other cells using essentially two forms of transmission: electrical and chemical. In the CNS of higher vertebrates, chemical neurotransmission predominates. This chapter and Chapter 4 considers some of the cellular processes that are involved in the synthesis and processing of molecules that are involved in chemical neurotransmission, including the transmitters themselves, their receptors, and second messenger systems.

Forms of Chemical Neurotransmission

It is useful to distinguish several forms of chemical neurotransmission (Fig. 3.1). Neurons can be thought of as part of a continuum of secretory cells that communicate with other cells through the release of specific substances. At one end of this continuum are endocrine cells that do not have the highly differentiated form of neurons; these cells release their hormones into the general circulation. At the other end of the continuum are neurons that target the release of their neurotransmitters to a highly specific site on another cell (the site of contact between the presynaptic terminal and the postsynaptic neuron). Intermediate between these extremes are *neurohormonal systems;* these specialized neurons in the hypothalamus give rise to axons that terminate on blood vessels. Neurons in the supraoptic and paraventricular nuclei of the hypothalamus send their axons into the neurohypophysis where they release oxytocin and vasopressin into the general circulation. Similarly, neurons in the tuberal nuclei of the hypothalamus terminate on "portal vessels" at the base of the pituitary gland. These neurons produce *releasing factors,* or "releasing hormones," that are conveyed via the hypothalamo-hypophyseal portal vessels to the anterior pituitary, where they induce the release of their particular hormones. Although the cells comprising the neurohormonal systems release hormones into the circulation, they are nevertheless neu-

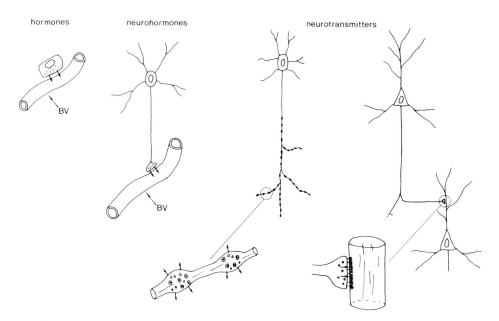

FIGURE 3.1. Hormones are released into the circulation (BV = blood vessel) by endocrine cells. Neurohormones are released into the circulation at specialized endings formed by neurons upon blood vessels. Neurotransmitters are released into the extracellular space (neuromodulators) or directly on a postsynaptic membrane specialization of another cell.

rons because of their lineage, because they elaborate characteristic axons and dendrites, and because they also form traditional synapses upon other neurons within the CNS.

A few types of neurons may also give rise to terminals that end "blindly" near other neurons rather than forming traditional synapses. Such terminals appear to release their transmitters into a local cellular environment rather than upon a specific postsynaptic site. Thus, the targeting of release for these neurons is slightly less specific than in the case of traditional synapses, but more specific than neurohormonal systems. This mode of termination has been suggested for serotoninergic systems.

Types of Receptors

It is also useful to distinguish between *ionophore-linked* and *second messenger-linked* receptors. Ionophore-linked receptors are joined with or are part of an ion channel that opens in response to the neurotransmitter. The opening of the channel causes a rapid conductance change, which typically is short-lasting. Second messenger-linked receptors are coupled to enzymes that produce intracellular *second messengers* (cAMP and the products of phosphatidyl inositol degradation, for example); these second messengers may produce long-lasting metabolic effects in their postsynaptic targets, usually by regulating the phosphorylation of some other enzymes (see Chapter 4). The second messenger-linked systems are also referred to as *neuromodulators*. Some of the second messenger-linked systems exhibit the intermediate form of termination described above, in which the nerve terminals release their substances into a local cellular environment rather than onto a particular postsynaptic site.

In making the distinction between ionophore-linked and second messenger-linked neurotransmitters, it is important to note that ionophore-linked systems may also produce a second messenger. For example, it is thought that one of the receptors for the excitatory amino acids is linked to a calcium channel (the *N*-methyl-D-aspartate receptor, see below). The opening of this channel is thought to result in increases in intracellular calcium, which serves a second messenger function to modulate Ca^{2+}/calmodulin-dependent processes.

Criteria for Identifying a Substance as a Neurotransmitter

Much effort has been expended to identify chemical neurotransmitters since chemical transmission was first recognized. In studies of potential neurotransmitters, several criteria must be met in order to establish that the molecule plays a neurotransmitter role. These criteria are as follow:

1. Localization: The substance should be localized in nerve endings.
2. Release: The substance should be released when the terminals are activated.
3. Physiological identity: When applied to the postsynaptic site, the substance should mimic the action of the natural neurotransmitter that is released from the terminal.
4. Pharmacological identity: Pharmacological manipulations should have identical effects both on the action of the putative neurotransmitter and on the natural neurotransmitter that is released from the terminal.

Other evidence often used to support a neurotransmitter candidate includes the presence of high affinity uptake mechanisms for the substance in a population of terminals and the existence of "receptors" (ie, binding sites for the substance) in association with particular pathways. The degree to which various neurotransmitter candidates have met these criteria varies considerably.

The first neurotransmitter substances that were discovered and characterized were relatively small molecules. More recently, it has become clear that larger peptide molecules also play a neurotransmitter role. Because of the different cellular mechanisms involved in processing (see below), it is convenient to divide neurotransmitters into *nonpeptide neurotransmitters* and *peptide neurotransmitters*. Nonpeptide neurotransmitters are discussed first.

Nonpeptide Neurotransmitters

Acetylcholine

Acetylcholine (ACh) has been recognized as a neurotransmitter since the 1920s, when it was found that ACh mimicked the action of the natural neurotransmitter in the heart. It is now recognized as the transmitter that is used at the neuromuscular junction, at ganglionic synapses in the sympathetic and parasympathetic nervous systems, and in the postganglionic fibers of the parasympathetic nervous system. Initial studies of the anatomical distribution of cholinergic pathways in the CNS took advantage of methods that revealed the location of the enzymes responsible for ACh degradation. Relatively simple histochemical methods were developed for the degradative enzyme acetylcholinesterase (AChE). By staining sections with this histochemical method, cell bodies, axons, and terminal fields of cholinergic neurons could be mapped (Fig. 3.2). More recent immunocytochemical studies using antibodies to the synthetic enzyme choline acetyltransferase (CHAT) have largely confirmed the early evidence derived from AChE histochemistry, although it has been found that AChE is occasionally present in some neurons that do not contain detectable quantities of CHAT. It is now thought that ACh is used as a neurotransmitter by several CNS pathways, particularly neurons in the caudate nucleus, basal forebrain, and brainstem.

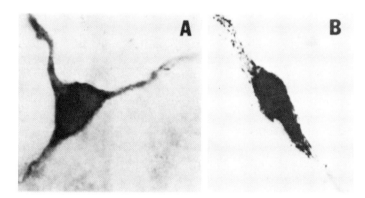

FIGURE 3.2. Identifying neurons that use particular transmitter substances by their content of neurotransmitter-related enzymes. *A* Cholinergic neurons in the caudate nucleus stained using antibodies to CHAT. *B* Cholinergic neurons in the caudate nucleus stained for AChE using a histochemical method. [Reprinted with permission from Wainer BH et al. (1984) Cholinergic systems in mammalian brain identified with antibodies against choline acetyltransferase. *Neurochem Int* 6: 163–182, Copyright 1984, Pergamon Press plc.]

Catecholamines and Indole Amines

Next to be recognized and characterized were the catecholamines *(nor-epinephrine, dopamine)* and the indole amine *serotonin* (5-hydroxytrypt-amine, 5-HT). Norepinephrine is accepted as the neurotransmitter in the postganglionic fibers of the sympathetic nervous system. The anatomical distribution of catecholamine and indoleamine pathways in the CNS has been defined by methods that reveal the presence of the neurotransmitter itself. Of particular importance were the histofluorescent procedures developed initially by Falck and Hillarp (Falk et al. 1962). These sensitive methods have enabled a reasonably complete characterization of the cells of origin of the various neurotransmitter substances, as well as of the axons and terminal fields. These methods revealed that norepinephrine is present in high concentrations in certain nuclei in the brainstem, particularly the neurons in the locus ceruleus. Dopamine is present in neurons of the nigrostriatal pathway that projects from the substantia nigra to the caudate-putamen complex and also in neurons in several small nuclei in the brainstem. Serotonin is present in high concentrations in neurons of the raphe nuclei of the brainstem. Thus, it is now thought that norepinephrine, dopamine, and serotonin are synthesized by neurons in rather small nuclear groupings in the brainstem that have wide-ranging projections to other brain regions.

γ-Aminobutyric Acid (GABA)

The amino acid derivative *GABA* is an inhibitory neurotransmitter substance in the vertebrate CNS. The anatomical distribution of GABAergic systems has been defined using antibodies against the enzyme responsible for GABA synthesis, glutamic acid decarboxylase (GAD). Both cell bodies and terminals of GABAergic pathways have been characterized, and it is clear that GABAergic neurons correspond to inhibitory interneurons in most brain regions. These GABAergic interneurons project locally and generally form basket-type endings around the perikarya of other neurons. GABA-containing synapses also terminate upon axon initial segments and upon dendritic shafts. Cerebellar Purkinje cells are also thought to use GABA as their neurotransmitter. In most CNS locations the GABAergic synapses form type II (symmetric) synapses (see Fig. 3.3). Under appropriate fixation conditions, these synapses have *flattened vesicles* rather than the round vesicles that are characteristic of excitatory synapses. It is now thought that GABA is the principal inhibitory neurotransmitter in the CNS.

Amino Acids

The amino acids *glutamate and/or aspartate* have been implicated as neurotransmitters in a number of very prominent CNS pathways. Unfortu-

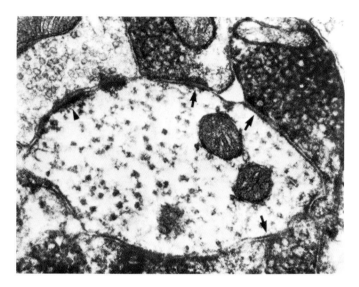

FIGURE 3.3. Electron micrograph of the substantia nigra after immunocytochemistry using an anti-GAD serum. Terminals filled with GAD-positive reaction product form symmetric synapses with a dendritic shaft. An unstained terminal that contains round synaptic vesicles forms an asymmetric synapse nearby. ×67,000 [From Ribak CE et al. (1976) Immunocytochemical localization of glutamate decarboxylase in rat substantia nigra. *Brain Res* 116: 287–298.]

nately, however, the properties of other neurotransmitter substances that have provided a means for experimental investigation are not available with the excitatory amino acids. For example, there is as yet no strong evidence for unique enzymes for the synthesis of the excitatory amino acids that serve a neurotransmitter function; as far as is known, the excitatory amino acid neurotransmitters are derived from metabolic pathways that are present in all cells (the Krebs cycle). Thus, it has not been feasible to develop antibodies to unique synthetic enzymes for use in immunocytochemical studies. Moreover, the excitatory amino acids themselves are not uniquely localized in neurons that use them as transmitters; a crucial problem has been distinguishing neurotransmitter pools from metabolic pools. Thus, attempts to define the pathways that use the excitatory neurotransmitters by the presence of the substances themselves have met with little success.

Despite the problems of experimental investigation, evidence of a neurotransmitter role for the excitatory amino acids in certain CNS pathways is reasonably strong. The evidence is as follows:

1. Certain populations of synaptic terminals contain high concentrations of the amino acids, as demonstrated by the fact that the endogenous concentration of the amino acids is reduced following lesions.

2. The same population of synaptic terminals have selective high affinity uptake systems for aspartate and glutamate.
3. The excitatory amino acids are released as a result of physiological activation of the pathways.
4. The excitatory amino acids meet the physiological and pharmacological identity criteria, that is, when applied iontophoretically, the amino acids have physiological and pharmacological properties that are similar to the natural neurotransmitter.
5. Binding studies suggest the presence of specific excitatory amino acid receptors in association with the pathways (see below).

The evidence is often confusing whether it is glutamate or aspartate that is the more likely neurotransmitter candidate. Part of the problem is that the high affinity uptake systems are very similar. In addition, destruction of pathways sometimes leads to parallel changes in both glutamate and aspartate. In some systems of glutamate and aspartate may actually operate as cotransmitters. It has also been suggested that a dipeptide consisting of both molecules is the actual neurotransmitter. Hopefully, future research will resolve some of these ambiguities.

For many pathways when the evidence does favor one amino acid over the other, glutamate is the favored candidate. Thus, the most likely neurotransmitter candidate in a variety of cortical projection systems is glutamate. The best evidence of a pathway that might use aspartate involves the commissural pathway of the hippocampus. In any case, glutamate and aspartate are now thought to be the most common excitatory neurotransmitter for CNS neurons, particularly those in cortical structures.

The amino acid *glycine* is thought to operate as an inhibitory neurotransmitter for some neurons in the spinal cord, medulla, and pons. The evidence suggesting a neurotransmitter function is (1) the high endogenous concentration of the amino acid, (2) the presence of high affinity uptake systems, (3) the fact that the amino acid is released from terminals, and (4) the presence of receptors (in this case, strychnine-binding sites). There is also limited evidence that glycine has physiological and pharmacological properties similar to the natural neurotransmitter. Studies of the uptake of radiolabeled glycine and GABA reveal that different populations of neurons accumulate each molecule. For this reason it is thought that glycine and GABA may be used by separate populations of inhibitory interneurons.

Other Neurotransmitter Candidates

In addition to the molecules described above, a number of other molecules have been proposed as possible neurotransmitters, including adenosine, epinephrine, and histamine. These substances are unquestionably "neuroactive" in that they affect neural activity, but less information is available concerning the neurotransmitter role of these compounds.

TABLE 3.1. Neuropeptides.

Neuropeptide	Abbreviation	No. of AA residues	Peptide Family
Adrenocorticotropic hormone	ACTH	24	POMC
Amidorphin	AMD	26	Proenkephalin
Angiotensin II	AII	8	Renin
Atrial natriuretic factor	ANF	126	Atriopeptides
Bombesin	BO	14, 27, 32	BO/GRP
Bradykinin	BK	9	Kinin
Calcitonin gene-related peptide	CGRP	37	Pro-CGRP
C Fragment	LPH61-87	27	POMC
Cholecystokinin	CCK	3, 4, 8, 12, 33, 39	Gastrin/CCK
CLIP (ACTH18-29)	CLIP	12	POMC
Corticotropin-releasing factor	CRF	41	
Dynorphin	DYN	8, 17	Prodynorphin
α-Endorphin	α-End	15, 16	POMC
β-Endorphin	β-End	31	POMC
γ-Endorphin	γ-End	16, 17	POMC
Methionine-enkephalin	met-Enk	5	Proenkephalin
Leucine-enkephalin	leu-Enk	5	Proenkephalin
FMRF-amide		4	
Gastrin			Gastrin/CCK
Gastrin-inhibiting peptide	GIP	43	GRF/PHI
Gastrin-releasing peptide	GRP	27	BO/GRP
Glucagon	GLU	29	GRF/PHI
Growth-hormone-releasing factor	GRF, GHRH	44	GRF/PHI
Leumorphin	LUM	29	Dynorphin
β-Lipotropin	β-LPH	91	POMC
Luteinizing-hormone-releasing hormone	LHRH, LRF	10	
α-Melanocyte-stimulating hormone (ACTH1-13)	α-MSH	13	POMC
β-Melanocyte-stimulating hormone (LPH41-58)	β-MSH	18	POMC
γ-Melanocyte-stimulating hormone	γ-MSH	11, 12, 27	POMC
Metorphamide (adrenorphin)	MOA	8	Proenkephalin
Motilin	MO, PIM	22	Tachykinin
Neuromedin B	NMD-D	10	BO/GRP
Neuromedin C	NMD-C	10	BO/GRP
Neuromedin K (neurokinin-β)	NMD-K, β-NK	9	Tachykinin
Neuromedin L (substance K)	NMD L	9	Tachykinin
Neuropeptide Y	NPY	36	NPY/PP
Neurotensin	NT	13	
Oxytocin	OXY	9	Prooxytocin
Peptide E	PEP-E		Proenkephalin
Porcine intestinal peptide	PHI	27	GRF/PHI
Ranatensin	RT	11	BO/GRP
Secretin	SEC	27	GRF/PHI
Somatostatin*	SOM	13, 14, 28	Prosomatostatin
Substance K (see NMD-L)	SK	9	Tachykinin

TABLE 3.1 (Continued)

Neuropeptide	Abbreviation	No. of AA residues	Peptide Family
Substance P	SP	11	Tachykinin
Thyrotropin-releasing hormone	TRH, TRF	3	
Vasoactive intestinal polypeptide	VIP	28	GRF/PHI
Vasopressin	AVP	9	Provasopressin

*Growth hormone release inhibiting factor
POMC = pro-opiomelanocortin; PP = pancreatic polypeptides; PHI = porcine intestinal peptide; BO = bombesin

Neuropeptides

A large class of neurotransmitters are the *neuropeptides;* these are small peptide molecules containing from 3 to about 100 amino acid residues (see Table 3.1). Among the neuropeptides are the "neurohormones," including the hypothalamic-releasing factors (Fig. 3.4); indeed, oxytocin and vasopressin were among the first of the neuropeptides that were recognized and characterized. During the past decade, a large number of other neuropeptides have been identified that are thought to play a role as neurotransmitter or neuromodulator substances. The development of specific antibodies to the individual neuropeptides has provided an important means to define their anatomical distribution using immunocytochemistry. The hypothalamus is particularly rich in neurons that contain neuropeptides. However, neuropeptides are also found in all other parts of the nervous system. In cortical structures, the neuropeptide-containing cells often seem to be a class of interneuron.

A particularly interesting subgroup of neuropeptides are the endogenous opiatelike substances. These substances have in common an *opioid core* consisting of the amino acid sequence Tyr-Gly-Gly-Phe-Met (or Leu). The functions and physiological properties of the neuropeptides are not yet clearly defined, but there is evidence that they have modulatory roles. Some of the neuropeptides are capable of modulating the effect of other neurotransmitters. In this regard, it is of interest that neuropeptides are often co-localized with other neurotransmitters and thus may act as *cotransmitters* (see below).

Co-Localization

Recent studies of transmitter localization have revealed that individual synaptic terminals may contain more than one neurotransmitter. This is particularly true of the neuropeptide systems. Sometimes, neuropeptides are co-localized with other types of neurotransmitters such as GABA, ACh, dopamine, norepinephrine, or serotonin. In other situations, different

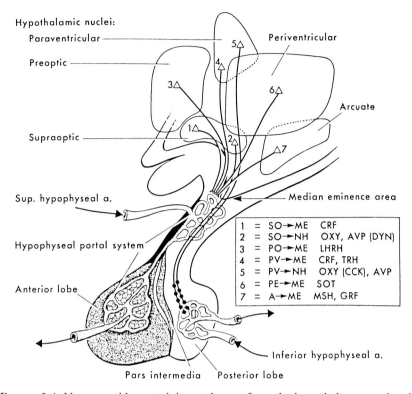

FIGURE 3.4. Neuropeptide-containing pathways from the hypothalamus to the pituitary. Projections from the paraventricular and supraoptic nucleus terminate in the posterior lobe of the pituitary and release oxytocin and vasopressin. The hypothalamic-releasing factors are contained in the pathways terminating in the median eminence. These factors are released into the hypophyseal portal system and carried to the anterior lobe where they regulate the release of their target hormones. [From McGeer PL, Eccles JC, McGeer EG (1987) *Molecular Neurobiology of the Mammalian Brain.* New York, Plenum Press.]

neuropeptides are co-localized, which implies that the substances may be released simultaneously as cotransmitters. Thus, mechanisms apparently exist to allow neurons to release a blend of neurotransmitter substances rather than a single substance; obviously, such blends considerably enrich the options available to neurons to affect their targets.

Cellular Mechanisms of Chemical Neurotransmission

As is evident from Table 3.1, neuroactive substances range from simple molecules of the size of amino acids to peptide molecules of varying complexity. These substances obviously have different characteristics and require different cellular mechanisms for their synthesis and processing. The amino acids, for example, need not be specially synthesized; they need only be accumulated at the appropriate site (the presynaptic terminals). Synthesis of the more simple neurotransmitters may depend on a single rate-limiting enzyme. Synthesis of the polypeptides requires translation of a specific mRNA and posttranslational processing that may require several additional enzymatic reactions. Before discussing these differences in detail, it is useful to consider the general activities that neurons must undertake in order to use chemical neurotransmitters.

For any neurotransmitter the presynaptic neuron must (1) *synthesize or accumulate* the neurotransmitter substance, (2) *deliver* the neurotransmitter to the release site (usually presynaptic terminals), (3) *store* the neurotransmitter substance in a form that is ready to be released (presumably in vesicles), and (4) *release* the neurotransmitter in response to the appropriate stimulus (usually an action potential invading the presynaptic terminal). Presynaptic neurons may also inactivate the neurotransmitter through reuptake and/or degradation, although the postsynaptic cell and glial cells may also play a role in uptake or degradation. The postsynaptic cells must also construct membrane sites associated with each presynaptic terminal; these sites must have the appropriate neurotransmitter receptors, channels, and second messenger systems for the neurotransmitters to exert their effects. Neurons have developed specialized cellular mechanisms for each of these activities, and different types of neurons accomplish the same tasks in different ways.

Cellular Mechanisms of Neurotransmitter Synthesis or Accumulation

Neurotransmitters may be synthesized from precursors if they are specialized molecules, or if they are generally available (the amino acids), they may be accumulated by terminals and concentrated. The synthesis of neurotransmitters may take place locally in the presynaptic terminals or in the cell body. The former method requires that the enzymes necessary for neurotransmitter synthesis be transported to the terminal from the cell body. The latter method requires that the neurotransmitters themselves be transported to the presynaptic terminal after synthesis.

Transmitters That Are Synthesized Locally in Synaptic Terminals

The nonpeptide neurotransmitter molecules such as ACh, GABA, norepinephrine, dopamine, and serotonin are locally synthesized within nerve terminals. For these neurotransmitters the critical synthetic enzymes are produced in the cell body and transported to the presynaptic terminals.

The enzymes responsible for neurotransmitter synthesis may be localized either in the nerve terminal cytoplasm or in vesicles. For example, CHAT is localized in the terminal cytoplasm; after synthesis, ACh is rapidly concentrated in synaptic vesicles (Fig. 3.6). The synthetic enzymes for the catecholamines are localized in *chromaffin granules* (in the case of the adrenal medulla) or in dense core vesicles for noradrenergic and dopaminergic neurons.

The presence of synthetic enzymes in terminals provides an important means for identifying the synapses associated with particular neurotransmitter systems. For example, it is possible to use antibodies for the enzymes in immunocytochemical procedures; synapses that contain the antigen can be readily identified at the light or electron microscopic level (Fig. 3.3). Such methods have been particularly useful for defining the distribution of ACh, norepinephrine, and GABAergic pathways.

Some of the precursors of locally synthesized neurotransmitters are available from intracellular pools, ie, acetylcoenzyme A (acetyl-CoA) for the acetate moity of ACh and glutamate for GABA synthesis. Other precursors are taken up by the nerve terminals by selective high-affinity uptake mechanisms. For example, cholinergic nerve terminals have a high-affinity uptake mechanism for choline; this uptake mechanism is selectively blocked by hemicholinium. The high-affinity choline uptake provides a marker for cholinergic terminals. Thus, subcellular fractions of nerve endings *(synaptosomes)* exhibit high-affinity choline uptake proportional to the number of cholinergic nerve terminals that are present. When the cholinergic innervation of a particular brain region is destroyed, the high-affinity choline uptake disappears coincident with the loss of the cholinergic terminals.

Although synthetic enzymes are present in terminals, and neurotransmitters are synthesized there, there also may be high concentrations of neurotransmitters throughout the cytoplasm of some neurons. As noted above, this fact has provided an important means to trace neurotransmitter pathways in the CNS, making it possible to define not only the terminal arborizations, but also the cells of origin.

Transmitters That Are Synthesized in the Cell Body

As would be expected, given that axons possess little capacity for protein synthesis, neuropeptide precursor proteins are synthesized in the neuronal cell body. The neuropeptides are derived from larger peptide precursor molecules by posttranslational processing. The initial neuropeptide gene product is a *pre-pro-protein*. The "pre-" sequence is the recognition sequence that directs the pre-pro-protein through the rough endoplasmic reticulum (RER) membrane. This pre-sequence is cleaved cotranslationally prior to the completion of synthesis, so that the *pro-protein* comes to reside in the cisternae of the RER. The pro-protein is then translocated from the RER cisternae into the Golgi apparatus, where it is packaged into *secretory granules* (Fig. 3.5).

Each of the biologically active neuropeptides is derived from pro-proteins by sequential proteolysis, sometimes followed by other posttranslational modifications such as acetylation. Thus, *pro-opiomelanocortin (POMC)* contains the sequences for the opioid peptide β-endorphin, as well as the sequences for adrenocorticotropic hormone (ACTH), β-lipotropin, and α-, β-, and γ- melanocyte-stimulating hormone (MSH). Similarly, *proenkephalin* is the precursor for met-enkephalin, leu-enkephalin, met-enkephalin-arg-phe, and met-enkephalin-arg-gly-leu, and *prodynorphin* is the precursor for α- and β- neo-endorphins, dynorphin A, and dynorphin B.

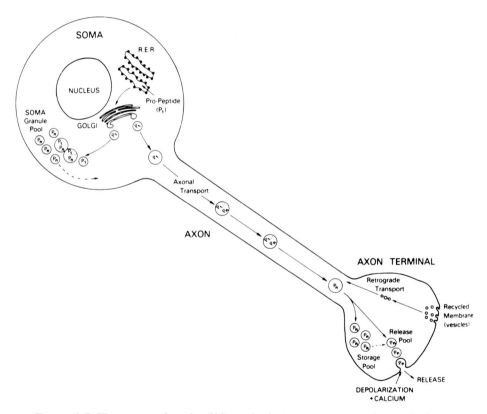

FIGURE 3.5. The presumed mode of biosynthesis, transport, processing, and release of peptides in a peptidergic neuron. Synthesis of the pro-peptide (P_1) occurs on the rough endoplasmic reticulum (RER) in the neuron soma. The pro-peptide is packaged into secretory granules in the Golgi apparatus, and posttranslational processing of the pro-peptide occurs in the secretory granules. Posttranslational processing can occur either in the neuronal soma or in the axon during transport. [Reproduced from *The Journal of Cell Biology* 1977; 73: 366–381, by copyright permission of the Rockefeller University Press.]

While it is clear that the pro-proteins are synthesized within the neuronal cell body, the site of posttranslational processing is known for only a few of the neuropeptides. For example pro-oxytocin and pro-vasopressin are synthesized in neuronal cell bodies in the hypothalamus and are progressively cleaved to oxytocin and vasopressin in secretory vesicles. Some of the processing occurs as the vesicles are transported down the axons (Fig. 3.5). Thus, Gainer and colleagues have proposed the *secretory vesicle hypothesis for precursor processing* (Gainer et al. 1977). Processing of other neuropeptides also seems to occur within secretory granules; apparently this processing takes place in the cell body and within axons. Since cleavage of several neuropeptides takes place in the secretory granules, the appropriate proteolytic enzymes must be present within the granules. Thus, processing could theoretically occur at any time after the granule is formed, including within the terminal. In this way, the production of the final neuroactive substance could be regulated within the terminal in essentially the same manner as other neurotransmitters.

Obviously, production of the neuropeptides depends not only on the mRNA for the precursors, but also on the production and positioning of the appropriate enzymes for posttranslational processing. By regulating how the pro-proteins are processed, neurons can control not only the amount, but also the blend of neuropeptides available in the secretory compartment (the secretory granules). If posttranslational processing occurs in the secretory granules, then the types of neuropeptides can be determined by regulating which processing enzymes come to be localized in the granules. Moreover, if the processing enzymes are localized in the granules themselves, an opportunity is provided for local regulation of the processing. The cellular mechanisms that underlie posttranslational processing are being actively explored by a number of laboratories.

Accumulating Amino Acid Neurotransmitters in Terminals

Amino acids are readily available to neurons in intracellular pools, and there is no definite evidence that special enzymes exist for producing the amino acids in neurons that use them as neurotransmitters. Nevertheless, the neurons that use amino acids as neurotransmitters must somehow generate a releasable pool of the amino acids. It is thought that neurons concentrate the amino acids in terminals, perhaps within vesicles. It is not certain exactly how this is accomplished.

One possible way that amino acid neurotransmitters could be concentrated in terminals is through the selective high-affinity uptake systems noted above. Presumably, these *transporters* are located in the external membrane of the terminal and are thus capable of taking up the amino acids from the extracellular fluid. High-affinity uptake is important in the removal of the transmitter from the junctional region after release (see below); the method has the added advantage that the transmitter can be reutilized.

Certainly, reuptake mechanisms are important for maintaining a releasable pool of transmitter under conditions of periodic release. The extent to which high-affinity uptake contributes to initial accumulation of the amino acid in the releasable pool in the terminal is less clear. Studies of glutamate turnover reveal that the specific activity of released glutamate is higher than the specific activity of the general glutamate pool. These results indicate that there is a preferential release of newly synthesized amino acid during neurotransmission, suggesting that the transmitter pool is derived from new synthesis rather than uptake. Obviously, our understanding of how the neurotransmitter pools of amino acid are regulated is sketchy.

Delivery of Neurotransmitters or Their Enzymes and Storage Within Terminals

Delivery

For neurotransmitters that are accumulated within terminals (the amino acids), or for transmitters that are synthesized locally within terminals, delivery is not a problem. However, the molecules that make up the high-affinity uptake systems or the synthetic enzymes must be delivered from the neuronal cell body. Little is known about the molecules of the high-affinity uptake systems, but since they are membrane proteins, it is probable that they are conveyed via the rapid axonal transport systems. Similarly, the synthetic enzymes that are localized in vesicles are usually conveyed via rapid transport systems.

Storage

It is thought that the primary storage depot for neurotransmitters is the vesicles. There is abundant evidence to support this contention for the catecholaminergic and cholinergic systems and for some peptides. The evidence is less clear for the amino acids. Studies of vesicle properties and vesicle storage mechanisms have been greatly aided by the development of subcellular fractions that permit the isolation of highly purified fractions of vesicles (Fig. 3.6).

Catecholamines are stored in *chromaffin granules* in the adrenal medulla and in *dense-core vesicles* in catecholaminergic neurons. As noted above, the synthetic enzymes tyrosine hydroxylase or dopa decarboxylase are also present within the vesicles. Other proteins, including *chromogranin,* are also present, along with high concentrations of adenosine triphosphate (ATP). At one time it was thought that the transmitter was stored in a stable 4:1 complex with ATP, but this is now believed to be incorrect, since the actual ratio of transmitter to ATP now seems to be much higher than 4:1 in most vesicles. The contents of the vesicles, including the syn-

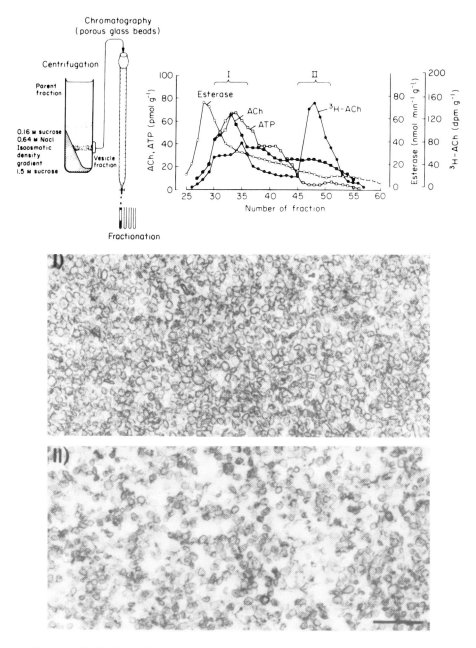

FIGURE 3.6. Evidence for localization of neurotransmitters in synaptic vesicles. Synaptic vesicles were isolated by subcellular fractionation and permeation chromatography. Large cholinesterase-positive particles are eluted first; these contain little ACh. ACh appears in two vesicle fractions (I and II; see electron micrographs of each fraction in the lower portion of the figure). When recently synthesized, ACh was labeled by incubating terminals with 3H choline or acetate; the labeled

thetic enzymes, are coreleased with the transmitter. Neuropeptides of various sorts are also present in the vesicles.

Cholinergic vesicles are much simpler in terms of their contents than the catecholaminergic storage vesicles. The principal contents are ACh (estimated to be about 880 mmol); ATP, and calcium. The vesicles contain essentially no soluble proteins, since CHAT is a cytoplasmic enzyme. The protein composition of the vesicle membrane is also quite simple; about five major polypeptides are present. The membrane does not appear to contain AChE or choline permease (the molecule responsible for high-affinity choline uptake), which are markers of the synaptic plasma membrane. This is of interest, since vesicle membrane is thought to be recycled during release by fusing with the plasma membrane and later being retrieved. Presumably, during retrieval some proteins of the synaptic plasma membrane are excluded from the vesicle membrane. ATPase is present in the vesicle membrane, as is actin.

The nature of the storage depot for the amino acid neurotransmitters is less clear. Presumably these are stored in vesicles, but the possibility that the amino acids are stored in some cytoplasmic pool cannot be excluded. If the amino acids are stored in vesicles, little is known about the contents of the vesicles, or the mode of storage.

Release (Stimulus-Coupled Secretion)

The *vesicle hypothesis* of neurotransmission holds that neurotransmitters are stored in terminals in vesicles and that release involves *exocytosis* whereby these vesicles fuse with the terminal's external membrane so that the lumen of the vesicle opens into the synaptic cleft. During high-frequency activation, so-called *omega profiles* can frequently be observed, suggesting a fusion process (Fig. 3.7).

Quantitative studies by J. Heuser and T. Reese (1973) revealed that vesicles are lost after high-frequency activation of terminals and that there is an increase in the plasma membrane of the terminals that corresponds to the amount of membrane that would have been present in the vesicles that disappeared (Fig. 3.8). During recovery from high-frequency activation, the number of vesicles increases, and the amount of plasma membrane of the terminal decreases. These data strongly support the *vesicle recycling hypothesis* that vesicles fuse with the plasma membrane and

---

ACh was found preferentially in fraction II, which has smaller and denser vesicles than fraction I. [From Zimmerman H, Stadler H, Whittaker VP (1981) Structure and function of cholinergic synaptic vesicles, in Stajarne L, Hedqvist P, Lagercrantz H et al. (eds) *Chemical Neurotransmission, 75 Years.* New York, Academic Press.]

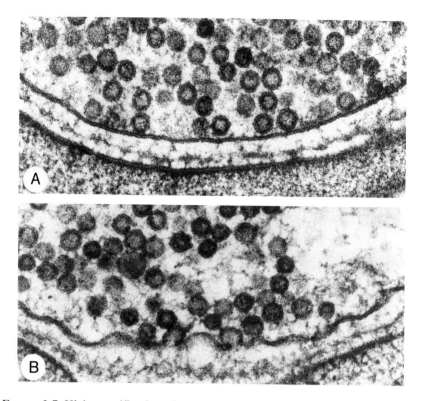

FIGURE 3.7. High magnification of nerve terminals at the frog neuromuscular junction: *A* at rest; *B* after stimulation in solutions of 4-aminopyridine, a treatment that greatly enhances release. The neuromuscular junctions were quick-frozen within seconds after stimulation and prepared for electron microscopy by freeze-substitution. Note the vesicles in *B* that appear to have fused with the plasma membrane. [Reproduced from *The Journal of Cell Biology* 1981; 88: 564–580, by copyright permission of the Rockefeller University Press. Micrograph produced by Dr. John E. Heuser of Washington University School of Medicine, St. Louis, MO.]

that this membrane is then recovered during the reformation of the vesicles (Fig. 3.9).

To affect the appropriate site, vesicles must be released into the synaptic cleft. Since the entire terminal is depolarized during the action potential, there must be a mechanism for directing the vesicles to the appropriate site on the terminal membrane. Studies of presynaptic terminals at neuromuscular junctions have revealed that there are "active zones" that are characterized by arrays of intramembrous particles where vesicles fuse (Fig. 3.10). These may represent some sort of "docking site" for the vesicle.

Although the evidence supporting vesicular release is reasonably strong for cholinergic synapses, it is less compelling for other types of neuro-

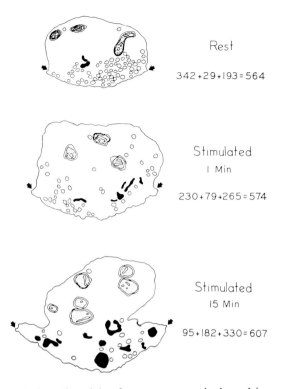

Rest

342+29+193=564

Stimulated

1 Min

230+79+265=574

Stimulated

15 Min

95+182+330=607

FIGURE 3.8. Depletion of vesicles from nerve terminals and increase in plasma membrane with prolonged stimulation. Motor axons were activated for 1 or 15 minutes at 10 Hz and examined with an electron microscope. The high-frequency stimulation led to a depletion of vesicles in the terminal and an increase in the surface area of the plasma membrane. [Reproduced from *The Journal of Cell Biology* 1973; 57: 315–344, by copyright permission of the Rockefeller University Press.]

transmitters. A viable alternative to the vesicular hypothesis is that the neurotransmitter may be released from a cytoplasmic pool even if the transmitter is stored in vesicles. Further work will be required to resolve this issue convincingly.

Calcium Dependency of Stimulus-Coupled Secretion

The release of neurotransmitter in response to depolarization (stimulus-secretion coupling) depends critically upon extracellular calcium. Elimination of extracellular calcium, or increases in extracellular magnesium decouple the excitation/secretion process. Depolarization of the terminal opens calcium channels in the plasma membrane; this leads to calcium influx into the nerve terminals, triggering the release of neurotransmitter

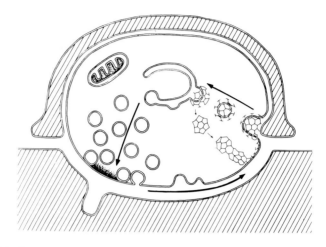

FIGURE 3.9. Summary of the vesicle recycling hypothesis. Synaptic vesicles release their contents by fusing with the plasma membrane at specific active zones. Equal amounts of membrane are retrieved by coated vesicles from regions of the plasma membrane adjacent to the Schwann sheath, and the coated vesicles lose their coats and coalesce to form cisternae. New vesicles are then derived from the cisternae. [Reproduced from *The Journal of Cell Biology* 1973; 57: 315–344, by copyright permission of the Rockefeller University Press.]

by presumably promoting the fusion of vesicles with the plasma membrane. Some of the characteristics of the calcium channels have been defined by studying the uptake of 45Ca by depolarized *synaptosome* preparations (subcellular fractions of pinched off nerve endings). The calcium channels that are present in synaptosomes are *voltage gated* (i.e., they open in response to changes in membrane potential) and appear similar to other voltage-gated calcium channels. The intramembranous particles that mark the active zones of synapses may actually be the calcium channels.

The extent of release from the terminal is determined by the duration of the calcium current, which, in turn, depends on the duration of the action potential. Action potential duration depends on the potassium conductance that terminates the action potential. A number of studies now reveal that this potassium conductance can be modulated. For example, in the neurons of the *Aplysia*, a marine gastropod, the release of neurotransmitter can be modulated by inactivation of the potassium conductance, which leads to an increase in action potential duration (see Chapter 4). This leads, in turn, to an increase in the amount of transmitter that is released. It is now thought that modulation of action potential duration may play an important role in modifying synaptic efficacy during behavioral plasticity.

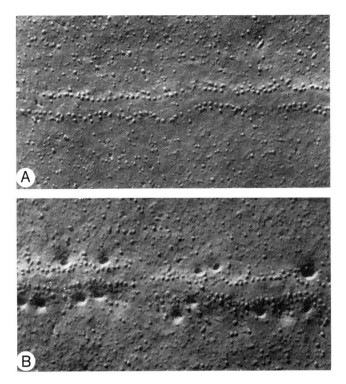

FIGURE 3.10. Freeze-fracture views of the plasma membrane of nerve terminals of the frog neuromuscular junction at rest (*A*) and after stimulation in 4-amino-pyridine (*B*). In freeze-fraction preparations, the plasma membrane is split along the lipid bilayer, permitting a view of either the inner or outer face of the plasma membrane. This view is of the inner face of the terminal membrane where the vesicles fuse. Proteins in the membrane appear as particles. As is evident in *A*, rows of particles are present in the terminal membrane. When the terminal is activated (*B*), "pits" appear along these rows of particles. The pits are actually views into the interior of vesicles that have fused with the membrane. The presence of particles at sites of vesicle fusion indicates that the terminal membrane has specialized regions for vesicle fusion. These are termed "active zones." [Reproduced from *The Journal of Cell Biology* 1981; 88: 564–580, by copyright permission of the Rockefeller University Press. Micrograph produced by Dr. John E. Heuser of Washington University School of Medicine, St. Louis, MO.]

Inactivation or Degradation of Neurotransmitters

After release, the action of neurotransmitters is terminated either by degrading the neurotransmitter molecule or by removing the transmitter from the synaptic cleft through uptake.

Acetylcholine is degraded by the enzyme *AChE*. The AChE at the neuromuscular junction is produced by the target cell (the muscle). For cholinergic neurons of the CNS that project intrinsically (ie, they do not give rise to an axon that projects peripherally), AChE is produced by the presynaptic neuron; the AChE is produced in the cell body and transported to the terminal. Thus, destruction of cholinergic terminals in the CNS leads to a loss of both CHAT and AChE from the target structure. As noted above, the presence of AChE in cell bodies provides a marker for cholinergic neurons of the central nervous system. At the terminal, AChE is associated with the plasma membrane, where it is capable of degrading the ACh that is present extracellularly.

Other nonpeptide neurotransmitters are removed from the synaptic cleft through uptake mechanisms. The synaptic terminal plays an important role in this process because of the presence of high-affinity *reuptake* mechanisms. However, glial cells also play a role in removing the transmitter, and glia possess enzymes that degrade neurotransmitter substances. The transmitter that is taken up by synaptic terminals can be reutilized; that which is taken up by glial cells is presumably degraded.

Much less is known about the inactivation of the peptide neurotransmitters. It is thought that neuropeptides are enzymatically inactivated by peptide hydrolysis in the synaptic cleft. However, direct evidence for this hypothesis is weak. Several enzymes have been identified that are capable of hydrolyzing neuropeptides, but it has not been established that these enzymes are actually present in the appropriate location (the synaptic cleft). Furthermore, in some cases the hydrolysis that does occur does not inactivate the peptides, but rather changes their biological activity. Thus, if neuropeptides are normally inactivated by enzymatic degradation, much is still to be learned about the mechanisms of this process. Still, there is little evidence for the alternative hypothesis that peptide action is terminated through reuptake mechanisms. The question of how neuropeptides are inactivated is likely to generate considerable research interest in the future.

Supplemental Reading

General

McGeer PL, Eccles JC, McGeer EG (1987) *Molecular Neurobiology of the Mammalian Brain.* New York, Plenum Press

Cooper JR, Bloom FE, Roth RH (1977) *The Biochemical Basis of Neuropharmacology,* ed 2. New York, Oxford University Press

Tracing Pathways that Use Specific Neurotransmitters

Falck B, Hillarp NA, Thieme G et al. (1962) Fluorescence of catecholamines and related compounds condensed with formaldehyde. *J Histochem Cytochem* 10:348–354

Shute CCD, Lewis PR (1963) Cholinesterase-containing systems of the brain of the rat. *Nature* 199:1160–1164

Endogenous Opioids

Akil H, Watson SJ, Young E et al (1984) Endogenous opioids: biology and function. *Ann Rev Neurosci* 7:223–255

Synthesis and Processing of Neuropeptides

Gainer H, Russel JT, Loh YP (1985) The enzymology and intracellular organization of peptide processing: the secretory vesicle hypothesis. *Neuroendocrinology* 40:171–184

Gainer H, Sarne Y, Brownstein MJ (1977) Biosynthesis and axonal transport of rat neurohypophyseal proteins and peptides. *J Cell Biol* 73:366–381

Loh YP, Brownstein MJ, Gainer H (1984) Proteolysis in neuropeptide processing and other neural functions. *Ann Rev Neurosci* 7:189–222

Release of Neurotransmitter

Heuser JE, Reese TS (1973) Evidence for recycling of synaptic vesicle membrane during transmitter release at the frog neuromuscular junction. *J Cell Biol* 57:315–344

Heuser JE, Reese TS (1981) Structural changes after transmitter release at the frog neuromuscular junction. *J Cell Biol* 88:564–580

Katz B (1969) *The Release of Neural Transmitter Substances.* Liverpool, England, Liverpool University Press

CHAPTER 4

Interneuronal Communication II: Neurotransmitter Receptors and Second Messenger Systems

Introduction

Neurotransmitters are not inherently excitatory or inhibitory; in fact, certain neurotransmitters are excitatory in some situations and inhibitory in others. The action of a neurotransmitter depends upon its postsynaptic receptor and the intracellular processes that are influenced by the receptor. As noted in Chapter 3, it is useful to distinguish between two general classes of receptors: those linked to or part of ion channels (ionophore-linked) and those coupled to second messenger-generating systems (second messenger-linked). These will be considered separately.

Ionophore-Linked Receptors

In general, excitatory neurotransmitters that are ionophore-linked open membrane channels that permit inward current flow, or close channels that produce a resting outward current flow. Inhibitory neurotransmitters that are ionophore-linked open membrane channels that permit outward current flow, or close channels that produce a resting inward current flow. Thus, the action of the neurotransmitter depends upon the ion channel associated with the receptor.

In principle, receptor and channel functions could be mediated either by a single protein molecule, or by separate proteins whereby the receptor regulates a separate channel protein. In fact, the receptor-ionophores that have been characterized are protein molecules that subserve both receptor and channel functions.

The ACh Receptor

The best understood receptor is the nicotinic ACh receptor. In response to the binding of ACh, a *cation-specific ion channel* is rapidly opened, leading to an increase in Na^+ and K^+ conductance. The increased conductance leads to a rapid depolarization of the membrane.

The rapid growth of knowledge about the ACh receptor depended upon (1) the existence of sources that were rich in receptors, such as the electric organ of *Torpedo,* the electric fish; and (2) the discovery of snake toxins that selectively and irreversibly bind to the receptor, such as alpha bungarotoxin from a Southeast Asian snake called the banded krait. Thus, it was possible to isolate the receptor to a high degree of purity and define its composition. Once the protein was purified, it was possible to sequence the proteins comprising the receptor, isolate the mRNAs coding for the protein, and prepare complementary DNAs (cDNAs) to the messages so that the molecular genetics of the protein could be defined.

The receptor protein was found to be a heavily glycosylated molecule consisting of subunits arranged as a pentamer (two alpha, one beta, one gamma, and one delta) with apparent molecular weights of 54,000 (54K), 56,000 (56K), 58,000 (58K), and 60,000 (60K). The subunit proteins comprising the ACh receptor are quite similar, having long stretches of homologous amino acid sequences. Each subunit is apparently encoded by a different gene. The protein has five stretches of amino acids in α-helix configurations that represent potential membrane-spanning domains. It is thought that each subunit has one such domain, and that each subunit spans the membrane. The ion channel is essentially a pore, the sides of which are formed by portions of the membrane-spanning domains of each of the five subunits (Fig. 4.1).

The ACh binding site on the molecule has been identified by applying cholinergic ligands that can be cross-linked to nearby amino acids. One ligand that competes with ACh reacts covalently with nearby sulfhydryl groups of amino acids upon treatment with a reducing agent. Another photoactivatable cholinergic ligand reacts covalently when irradiated with ultraviolet (UV) light. Peptide mapping and sequencing methods can then define where the crosslinking occurs in the protein molecule. Using such methods, the ACh recognition site has been localized to two cysteine residues on the two α-subunits.

A similar approach has permitted an identification of the portions of the molecule involved in channel functions; for these studies photoactivatable noncompetitive blockers have been employed. These agents block the permeability response induced by ACh without competing with cholinergic agonists for the binding sites. For example, when the channel blocker chlorpromazine is crosslinked by photoirradiation, it binds to sites on all subunits. This result suggests that chlorpromazine acts by plugging the channel and that when the chlorpromazine molecule is in this position, it is equidistant from each of the subunits. This is taken as evidence that the transmembrane portions of each subunit form the wall of the channel. Interestingly, chlorpromazine binding occurs with a much faster time course in the presence of ACh, suggesting that access of chlorpromazine to the binding site is facilitated when the channel opens. These data suggest a physical change in the receptor-ionophore complex when the channel opens.

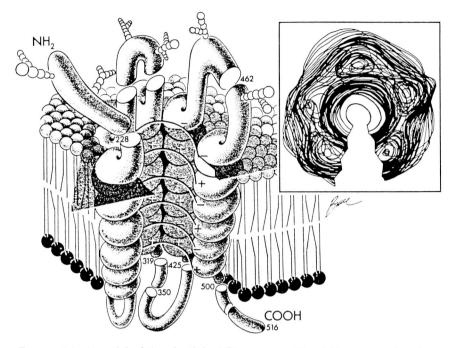

FIGURE 4.1. A model of the nicotinic ACh receptor. The ACh receptor has five membrane-spanning domains that are thought to correspond to the stretches that are in an α-helix configuration in each subunit. Cuts indicate where segments of the protein are not shown. [Adapted from Young EF et al. (1985) Topological mapping of acetylcholine receptor: Evidence for a model with five transmembrane segments and a cytoplasmic COOH-terminal peptide. *Proc Natl Acad Sci USA* 82: 626–630.] *Inset:* Three-dimensional contour map of the receptor based on electron microscopic images of purified receptors. [Reprinted by permission from *Nature* 315: 474–480. Copyright © 1985 Macmillan Magazines Ltd. Composite figure from McGeer PL, Eccles JC, McGeer EG (1987) *Molecular Neurobiology of the Mammalian Brain,* New York, Plenum Press.]

By cloning and sequencing cDNAs for each of the subunits comprising the receptor and then altering the cDNA using *site-directed mutagenesis,* it has been possible to further characterize how the hydrophobic portions of the subunits contribute to channel properties. Modified mRNA molecules produced by the altered cDNAs are injected into *Xenopus* oocytes, where they express the altered receptor. The properties of the receptor can then be evaluated physiologically by recording from the injected oocytes during the application of cholinergic ligands. Injections of unaltered mRNA lead to the expression of functional ACh receptors that have properties very similar to native receptors. However, injections of mRNA altered in particular locations lead to the expression of receptors with altered channel properties, thereby permitting a determination of the sequences that are crucial for channel functions.

Research on the ACh receptor is continuing at a rapid pace, and it is becoming possible to explain a variety of receptor-mediated functions at the molecular level. For example, a number of allosteric sites on the molecules have been identified, where particular ligands induce conformational changes in the molecule, altering channel properties. One example of such an allosteric transition is thought to occur during desensitization; the receptor is thought to undergo a transition between a resting activatable and an inactive conformation in response to prolonged exposure to ACh. Similar conformational changes may occur in response to regulatory ligands or second messengers that modify channel function. It is expected that many physiological modifications of channel function will be explainable on the basis of molecular properties.

GABA and Glycine Receptors

In response to the binding of their respective agonists, GABA and glycine receptors open *anion-specific membrane channels,* leading to an increase in chloride conductance. Purification of the receptors has been feasible, again because of the existence of well-characterized ligands; thus, the GABA receptor has been purified to homogeneity using benzodiazepine-affinity chromatography, and the glycine receptor has been purified using strychnine-affinity chromatography. Although these have not yet been as well characterized as the ACh receptor, the available data suggest that these receptors are similar in many ways to each other and to the ACh receptor.

The GABA receptor consists of two subunits (alpha and beta) with apparent molecular weights of 53K and 57K. Again, there are similar amino acid sequences in some parts of the two molecules. There is also considerable homology between the subunits of the GABA receptor and those of the ACh receptor. In particular, the subunits are of a similar size and possess similar sequences that are predicted to be the hydrophobic transmembrane regions of the proteins. There are also extracellular portions of the protein molecule that are very similar in the different receptor classes.

To develop cDNA probes for the subunits of the GABA receptor, the receptor was enzymatically cleaved, and the amino acid sequences of the fragments were determined. Complementary oligonucleotide probes were then constructed and used to screen cDNA libraries prepared from mammalian brain. The clones that code for the receptor subunits were identified and used to produce mRNA, which was then injected into *Xenopus* oocytes to confirm that these mRNA molecules coded for a functional receptor.

Using the *Xenopus* oocyte expression system, it has been possible to show that the α- and β-subunits are both necessary and sufficient for producing a GABA receptor coupled to a chloride channel. That is, injection of mRNA coding for both α- and β-subunits led to expression of the functional receptor, whereas injection of mRNA for either subunit alone did not lead to the expression of a functional receptor. The receptor-channel complex expressed in oocytes has properties that are similar to those of GABA receptors in brain.

The glycine receptor consists of three subunits with apparent molecular weights of 48K, 58K, and 93K. Fragments of the 48K protein have been sequenced, and corresponding oligonucleotide probes have been used to identify cDNA probes for the molecule. The sequences of the full copy cDNA probes were then determined, and the predicted amino acid sequence of the protein was deduced. Again, there were hydrophobic segments that were long enough to form transmembrane α-helices similar to those of the ACh receptor and the GABA receptor. There was also a considerable degree of sequence homology between the 48K subunit of the glycine receptor and the subunits of the ACh receptor.

Antibodies to the glycine receptor have now been developed, and they have been used in immunocytochemical studies: As would be expected, the receptor appears to be associated with the postsynaptic membrane specialization at type II (symmetrical) synapses, which often contain flattened vesicles (Fig. 4.2).

The structural similarities between the receptor-ion channel complexes

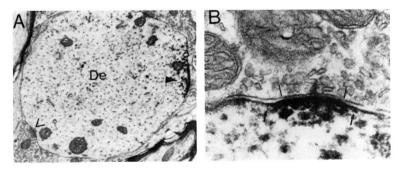

FIGURE 4.2. Localization of glycine receptors at synaptic sites on CNS neurons. Antibodies were prepared to glycine receptors (identified and purified on the basis of strychnine binding). The antibodies were used for electron microscopic immunocytochemical studies to localize receptors. *A* Dendrite (De) with two synaptic sites, one of which is stained with the antibody to the glycine receptor (*black arrowhead*). *B* Synaptic site positively stained for the glycine receptor. Note the flattened vesicles in the terminal that are typical of GABA or glycine-containing terminals. [Reproduced from *The Journal of Cell Biology* 1985; 101: 683–688, by copyright permission of the Rockefeller University Press.]

of the nicotinic ACh, GABA, and glycine receptors suggest that the different receptors may be part of a *"receptor super family"* that includes both anionic and cationic channels. This similarity makes sense because certain aspects of channel function are common. For example, the receptor-channel complex must span the membrane, the pore of the channel must be hydrophilic and permeable to ions, and the receptor must bind a ligand and induce a conformational change in the channel. Thus, it may be that various types of neurotransmitter receptors evolved from a common ancestor protein, perhaps one that served channel functions. That structurally similar channels are associated with binding sites for quite different receptors suggests that the two functions of the receptor-ion channel complex are discrete and separable. There is growing evidence for similar homology between receptors that are linked with second messenger-generating systems (see below).

Receptors for Excitatory Amino Acids

The excitatory amino acids also apparently use ionophore-linked receptors. Less is known about the molecular properties of these receptors, since it has not yet been possible to purify them. In large measure, the inability to purify the receptor is due to the absence of appropriate high-affinity ligands that can be used for receptor isolation. The key is a ligand that binds the receptor and not other proteins. What is known about excitatory amino acid receptors comes from pharmacological and electrophysiological studies of systems that are thought to use the excitatory amino acids. These studies have revealed that the excitatory amino acids use two types of receptor. One is similar to the ion channels described above in that binding of the neurotransmitter induces a rapid and relatively nonselective increase in conductance to cations and anions. The other type of receptor is quite different in that it is apparently a ligand-sensitive calcium channel whose function depends not only on ligand binding, but also on the membrane potential of the postsynaptic cell.

Fortunately, the different excitatory amino acid receptors have different agonist specificity. There are at least two types of receptors that open cation and anion channels; these receptors are activated by quisqualate and kainate, and are termed the quisqualate and kainate receptors, respectively. The receptor that opens a calcium channel is activated by N-methyl-D-aspartate (NMDA) and is thus called the NMDA receptor.

The most important feature of the NMDA receptor is that channel properties depend on the level of depolarization of the postsynaptic membrane (ie, channel function is *voltage-dependent*). The voltage dependency apparently comes about because the receptor is blocked in a voltage-dependent manner by magnesium ions; this blockade is relieved as the membrane is depolarized. Thus, the neurotransmitter will activate the channel only if the postsynaptic membrane is sufficiently depolarized. Usually,

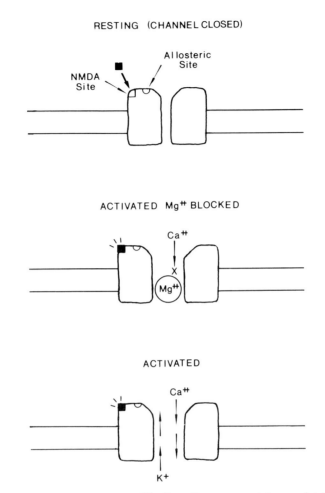

FIGURE 4.3. The NMDA receptor. Binding of an appropriate agonist (glutamate or NMDA) activates the receptor and opens the Ca^{2+} channel. However, with physiological concentrations of Mg^{2+}, the channel is blocked because of a cation-binding site within the channel. Thus, Mg^{2+} prevents Ca^{2+} flux. Mg^{2+} is displaced when the cell is depolarized for a sufficient period of time. Thus, the Ca^{2+} channel associated with the NMDA receptor typically opens only in response to prolonged presynaptic activity. The NMDA channel also has allosteric sites; glycine up-regulates receptor function, perhaps by facilitating agonist-induced transitions to the open channel configuration. [Reprinted by permission from *Nature* 329: 395–396. Copyright © 1987 Macmillan Magazines Ltd.

sufficient depolarization is achieved as a result of prolonged excitation. Thus, the excitatory amino acids may have two actions. The release of neurotransmitter in response to single presynaptic action potentials results in the activation of channels that depolarize the neuron. If the depolarization persists, magnesium is displaced from the NMDA receptor, thus opening a calcium channel that leads to a calcium current and to some calcium flux across the membrane (Fig. 4.3).

It is thought that the calcium that enters the neuron in response to the opening of the NMDA channel may serve as an intracellular second messenger (see below). In this way, the NMDA receptor is thought to function as both an ionophore-linked receptor and a second messenger-linked receptor.

Because the opening of the calcium channel associated with the NMDA receptor is voltage-dependent, the action of excitatory amino acid neurotransmitters released from a given synapse at any point in time depends on the history of activity over that synapse and over other synapses. Because of this property, the NMDA receptor is thought to play a role in mediating activity-dependent associative interactions between synapses. These interactions are thought to be important for inducing long-lasting changes in synaptic efficacy (e.g., long-term potentiation of synaptic efficacy, see below).

Like the other receptors, the NMDA receptor appears to be modulated through allosteric interactions. For example, phencyclidine and glycine lead to modifications in receptor function that do not apparently involve the NMDA binding site.

Second Messenger-Linked Receptors

The second major type of receptor is not linked to an ion channel, but is linked with enzymes that produce intracellular *second messengers*. These second messengers initiate a sequence of biochemical reactions that can lead to alterations in a number of cellular processes. For these receptors the neurotransmitter is considered a *first messenger*. Several second messengers have been identified, including cAMP, cyclic guanosine monophosphate (cGMP), calcium, and the breakdown products of phosphatidyl inositol, including inositol 1,4,5,-triphosphate (IP3) and diacylglycerol (DG).

cAMP as a Receptor-Modulated Second Messenger

A number of neurotransmitters—including dopamine, serotonin, norepinephrine, histamine, and octopamine—along with many hormones, activate the enzyme *adenylate cyclase* in their target cells. This activation occurs through membrane receptors, since it can be blocked by appropriate antagonists. The activation of the cyclase leads to the accumulation of *cAMP* within the cells, and the cAMP in turn stimulates the activity of *cAMP-dependent protein kinase,* leading to the phosphorylation of specific protein substrates (Fig. 4.4).

Studies of cAMP-linked receptor systems have revealed that receptors that stimulate second messenger production actually exert their effects through guanosine triphosphate (GTP) binding proteins (G proteins) that act as "go-betweens" for the receptor and the second messenger-generating systems. For example, G_s activates adenylate cyclase, whereas G_i inhibits it. Thus, the actions of particular neurotransmitters depend upon the G protein that is accessed by the particular receptor. The G proteins are also important for the other second messenger-generating systems discussed below, although the relevant G proteins are not as well-characterized as is the case for the cAMP system. Indeed, there may be a larger number of G proteins that link particular receptors and second messenger-generating systems.

cGMP as a Receptor-Modulated Second Messenger System

There are also neurotransmitter receptors that are coupled to cGMP-generating systems. Acetylcholine operating through muscarinic receptors, histamine, norepinephrine, and glutamate activate guanidylate cyclase through membrane receptors, leading to increases in intracellular levels

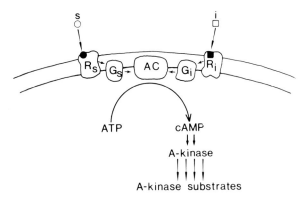

FIGURE 4.4. The cAMP second messenger system. Binding of the appropriate ligand to its membrane receptor leads to stimulation or inhibition of the enzyme adenylate cyclase (AC) that produces cAMP from ATP. Enzyme activation occurs through the action of a ligand on a `stimulatory receptor (R_s) operating through a G protein (G_s). Similarly, inhibition of the enzyme occurs via the action of a ligand on an inhibitory receptor (R_i) that operates in conjunction with a different G protein (G_i). Increases in intracellular cAMP activate cAMP-dependent protein kinase (A-kinase), which in turn phosphorylates a particular set of enzyme substrates within the cell.

of cGMP. Again, these effects are receptor-mediated, since they can be blocked by the appropriate antagonists. Cyclic GMP, in turn, activates a specific *cGMP-dependent protein kinase,* which phosphorylates a particular substrate protein termed the G-substrate. Cyclic GMP-dependent systems are particularly important in the cerebellum; the kinase is 10 to 50 times more concentrated in the cerebellum than in other brain regions, and the G-substrate is also about 50-fold more concentrated in the cerebellum. Immunocytochemical studies using antibodies to the protein kinase reveal that the enzyme is highly concentrated in Purkinje cells, and is undectable in most other cell types.

Phosphatidyl Inositol-Linked Second Messenger Systems

Another second messenger system that has been well characterized in studies of hormone action in a variety of cell types involves phosphatidyl inositol hydrolysis. In response to an agonist binding to its receptor, a phosphoinositidase (phospholipase C) is activated that cleaves membrane phosphatidyl inositol into inositol triphosphate or IP3 and DG. Both IP3 and DG then act as second messengers (Fig. 4.5). Inositol triphosphate is released from the membrane into the cytoplasm where it mobilizes intracellular calcium from nonmitochondrial stores (ie, it stimulates the re-

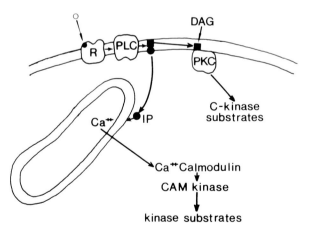

FIGURE 4.5. The phosphatidyl inositol (PI) second messenger system. Binding of the appropriate ligand to its membrane receptor stimulates phospholipase C (PLC), which hydrolyses membrane phosphotidyl inositol into diacylglycerol (DAG) and inositol phosphate (IP). DAG remains within the membrane and activates protein kinase C, which is a cytoplasmic enzyme that associates with the membrane. IP is released into the cytoplasm where it causes the release of Ca^{2+} from nonmitochondrial intracellular stores. The Ca^{2+} interacts with calmodulin to activate Ca^{2+} Calmodulin-dependent protein kinase (CAM kinase). Protein kinase C and CAM kinase phosphorylate a particular set of enzyme substrates.

lease of calcium from intracellular locations such as the endoplasmic reticulum where it is normally sequestered). The intracellular calcium then stimulates *calcium/calmodulin-dependent protein kinase*, resulting in the phosphorylation of the proteins that are normally the substrates of this kinase. Diacylglycerol remains in the membrane and activates *protein kinase C*, which phosphorylates a different set of substrate proteins. Thus, receptor-mediated hydrolysis of phosphatidyl inositol generates dual signals, each of which modulates different cellular processes.

Calcium as a Second Messenger

The DG produced upon hydrolysis of membrane phosphatidyl inositol is thought to exert its effect through the mobilization of intracellular calcium stores. In this situation calcium would actually be a *"third messenger"*; however, intracellular calcium can be regulated in other ways. For example, the NMDA receptor is thought to be a calcium channel that can, when opened, lead to increases in intracellular calcium. Similarly, voltage-dependent calcium channels in presynaptic terminals may bring about increases in calcium concentrations within the terminal after repetitive activation. Calcium can act as an intracellular second messenger in at least two ways (Fig. 4.6). First, the calcium can interact with calmodulin to stimulate *calcium/calmodulin-dependent protein kinase*. As will be seen below, there exist calcium/calmodulin-dependent kinases and appropriate

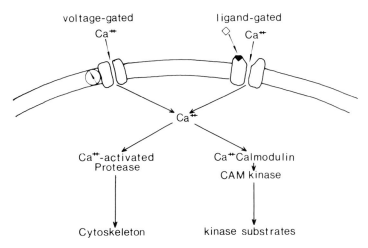

FIGURE 4.6. Ca^{2+} as a second messenger. Ca^{2+} can act as a second messenger in at least two ways: (1) by combining with calmodulin to activate CAM kinase and (2) by stimulating Ca^{2+}-activated proteases, which degrade components of the cytoskeleton. Intracellular Ca^{2+} concentrations can rise as a consequence of Ca^{2+} entry through voltage-activated Ca^{2+} channels, through ligand-gated channels (the NMDA receptor), or as a result of release from intracellular stores (see Fig. 4.5).

substrates for the kinases that might be modulated by increases in either postsynaptic or presynaptic calcium concentrations. Second, calcium can stimulate *calcium-activated protease,* an enzyme that selectively degrades certain elements of the cytoskeleton, particularly membrane-associated spectrin. The selective degradation of cytoskeletal proteins can then lead to a redistribution of membrane proteins including receptors. There is evidence that both types of calcium-mediated processes play an important role in regulating synaptic function.

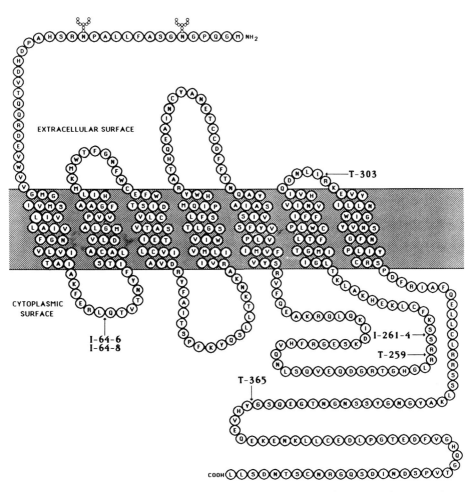

FIGURE 4.7. The amino acid sequence and presumed membrane architecture of the beta-2 adrenergic receptor. Note that there are seven membrane-spanning domains. Sites of mutations that are induced to study receptor function are illustrated by the *arrows.* [From Kobilka BK et al. (1987) Functional activity and regulation of human beta 2-adrenergic receptors expressed in *Xenopus* oocytes. *J Biol Chem* 262: 15796–15802.]

As is the case with the ionophore-linked receptors, rapid progress is being made in cloning the genes for second messenger-linked receptors. This has permitted the determination of the amino acid sequence of the receptors and the presumed membrane architecture (see Fig. 4.7 for the amino acid sequence and presumed membrane architecture of the human β-adrenergic receptor). The genes for the β-adrenergic receptor and two types of muscarinic cholinergic receptor have been cloned. All of these receptors have seven putative membrane-spanning domains, and all have sequence homology with one another. There is similar sequence homology between these receptors and the visual light receptor rhodopsin, which also activates a second messenger system. Thus, the second messenger-linked receptors may also be part of a receptor super-family.

The functional properties of second messenger-linked receptors can be studied by injecting the mRNAs for the receptors into oocytes, where the receptors are produced and incorporated into the oocyte membrane. The "induced" receptors become coupled with the second messenger-generating systems that are native to the oocyte. Using the oocyte system as an assay for receptor function, it is becoming possible to define the effects of mutations at specific locations on receptor function (Fig. 4.7). Using this strategy, it may be possible to define the functional domains on receptor molecules.

Functional Significance of Receptor-Mediated Second Messenger Systems

Each of the second messenger-linked systems can be thought of as part of a unique metabolic cascade. Binding of the agonist to the receptor modulates the second messenger-generating system (either stimulating or inhibiting it); the second messenger stimulates a specific kinase; and the kinase phosphorylates specific substrate molecules. Such systems are potential mechanisms for bringing about changes in neuronal signaling either by altering the electrophysiological properties of the membrane or by modulating synaptic transmission. In some situations there is evidence that the second messengers do, in fact, play such a role; indeed, for some of the second messenger-mediated cascades, the complete sequence of events from receptor activation to altered cellular function is known. For other cascades, only parts of the process have been defined. Indeed, the existence of certain cascades is only inferred from the presence of a receptor-mediated second messenger-generating system or from the existence of the second messenger-regulated kinases. Only a few of the cascades can be considered, but these examples indicate the broad range of possibilities for second messenger-mediated modulation.

cAMP and the Regulation of Synaptic Efficacy During a Simple Form of Learning in Aplysia

One of the most thoroughly characterized examples of an adenylate cyclase-mediated alteration in synaptic transmission involves the marine gastropod *Aplysia*. Studies by E. Kandel, J. Schwartz, and their associates defined the cellular processes that are involved in serotonin regulation of synaptic transmission over sensory pathways (for reviews, see Kandel et al. 1983; Schwartz et al. 1983). This example is of particular importance, since the behavioral significance of the synaptic modification is known.

Aplysia exhibit a simple form of learning termed *habituation* when presented with repetitive nonnoxious tactile stimuli. Initially, tactile stimulation elicits a withdrawl of the gill. If the tactile stimulation is presented repetitively, the gill-withdrawl response habituates (Figs. 4.8A, B). Kandel and colleagues defined in considerable detail the synaptic mechanisms of this simple form of learning and found that the decrease in the gill-withdrawl response occurs as the result of a decreased release of neurotransmitter by the terminals of sensory neurons (Fig. 4.8C).

Habituation occurs only when sensory stimulation is nonnoxious; when a noxious stimulus is delivered, there is a generalized increase in responsiveness termed *sensitization* (Fig. 4.8C). Again, the synaptic mechanisms of the increased behavioral responsiveness are known; the increased withdrawl response occurs because of an increased release of neurotransmitter of sensory neurons (Fig. 4.8).

The increase in synaptic efficacy during sensitization is brought about as a result of a serotoninergic input that is activated by the noxious stimulus. Serotonin exerts its effect by stimulating cAMP production (Fig. 4.9); cAMP then stimulates a protein kinase that phosphorylates a potassium channel, leading to channel inactivation. The inactivation of the channel slows the repolarization that normally occurs following the sodium-dependent portion of the action potential, leading to an increase in action-potential duration (Figs. 4.10 and 4.11). The end result is an increase in neurotransmitter release. Thus, a second messenger-linked receptor (for serotonin) modulates the function of an ion channel that, in turn, modulates neurotransmission.

Dopamine-activated cAMP Systems in the Central Nervous System

Another example of a cAMP-mediated alteration in cellular function involves dopamine. P. Greengard and his colleagues have defined how dopamine, acting through the so-called D1 receptor, stimulates the production of cAMP in neurons, which then leads to the phosphorylation of a protein phosphatase. D1 receptor binding sites and dopamine-sensitive adenylate cyclase activity are concentrated in regions that receive dopaminergic in-

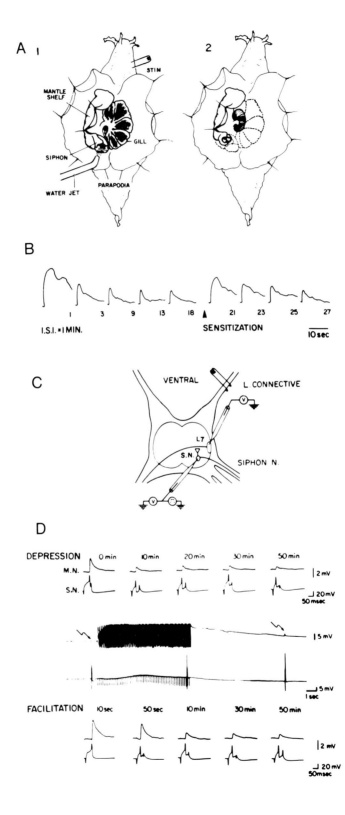

A 1

STIM

MANTLE
SHELF

GILL

SIPHON

WATER JET   PARAPODIA

2

B

1   3   9   13   18 ▲   21   23   25   27

I.S.I. = 1 MIN.   SENSITIZATION

10 sec

C

VENTRAL   L. CONNECTIVE

V

L 7

S.N.   SIPHON N.

V   C

D

DEPRESSION   0 min   10 min   20 min   30 min   50 min

M.N.

2 mV

S.N.

20 mV
50 msec

5 mV

5 mV
1 sec

FACILITATION   10 sec   50 sec   10 min   30 min   50 min

2 mV

20 mV
50 msec

108

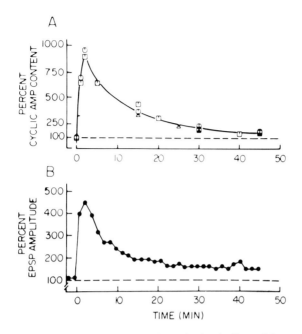

FIGURE 4.9. Increases in cAMP in the abdominal ganglion with sensitization. A sensitizing stimulus leads to increases in cAMP within the abdominal ganglion that occur concurrently with the increases in EPSP amplitude. *A* Time course of increase in cAMP. *B* Time course of sensitization as measured by increased EPSP amplitude. [From Kandel ER (1978) *A Cell-Biological Approach to Learning*. Grass Lecture Monograph, Society for Neuroscience, Washington D.C.]

FIGURE 4.8. Cellular mechanisms underlying short-term habituation and sensitization of the gill-withdrawal reflex in *Aplysia*. *A* Experimental arrangement for the behavioral studies. The gill-withdrawal reflex is elicited by a water jet to the siphon. The sensitizing stimulus is a noxious mechanical or electrical stimulus to the head. Habituating stimuli presented to the siphon at 1-min intervals lead to a decrease in responsiveness (the traces in *B* indicate the magnitude of the behavioral response). A sensitizing stimulus increases response amplitude. *C* Schematic illustrating simultaneous recording from gill motor neuron L7 (MN) and a mechanoreceptor sensory neuron (SN) in the abdominal ganglion of *Aplysia*. Repetitive stimulation of SN leads to a depression of the EPSP in MN; a sensitizing stimulus to the left connective, which carries information from the head to the abdominal ganglion, results in an increase in EPSP amplitude. These results reveal that habituation results from a decrease in synaptic efficacy at the SN-MN synapse, whereas sensitization results from an increase in synaptic efficacy at this same synapse. [From Kandel ER (1978) *A Cell-Biological Approach to Learning*. Grass Lecture Monograph, Society for Neuroscience, Washington DC.]

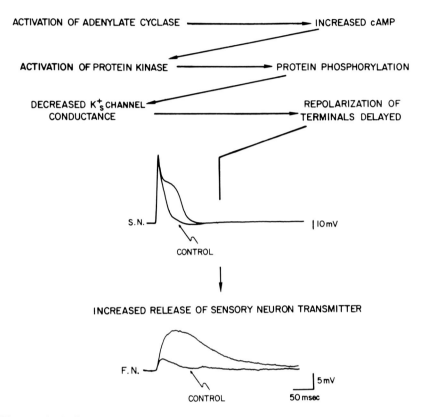

FIGURE 4.10. Summary of the events that are thought to underly presynaptic facilitation during short-term sensitization of the gill-withdrawal reflex. The sensitizing stimulus activates a group of facilitator neurons that release their transmitter onto the terminals of the sensory cells. Within the terminals, adenylate cyclase is activated, leading to the production of cAMP and the phosphorylation of A kinase substrates. One of the effects is a decrease in K^+ channel conductance, which leads to an increase in action-potential duration because the repolarization of the terminals is delayed. The traces from SN illustrate the results from an experiment in which the catalytic subunit of cAMP-dependent protein kinase was injected into the sensory cell. The result is a broadening of the spike in SN, which leads to a large increase in the EPSP evoked in the follower neuron (FN). [From Schwartz JH et al. (1983) What molecular steps determine the time course of the memory for short-term sensitization in *Aplysia*. *Cold Spring Harbor Symp Quant Biol* 47: 811–819.]

nervation (e.g., the caudate putamen complex and the ventral striatum). Dopamine activates a receptor that leads to an increase in cAMP in the neurons with the D1 receptors. This cAMP activates a cAMP-dependent protein kinase, which phosphorylates a 32 K molecular weight protein called DARPP-32, because it is *d*opamine and *c*AMP-*r*egulated *p*hospho-

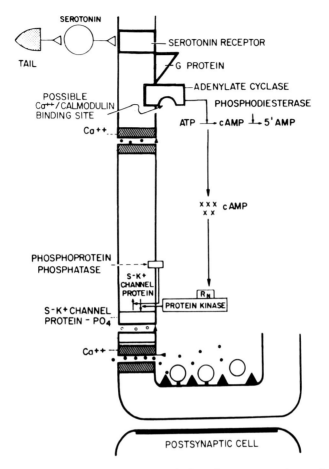

FIGURE 4.11. Cyclic AMP-dependent modulation of neurotransmitter release from presynaptic terminals in *Aplysia*. Serotonin, acting through its receptor and a G protein, activates adenylate cyclase, which leads to an increase in intracellular cAMP. The cAMP activates A kinase, which phosphorylates the K^+ channel protein and inactivates the channel, leading to a delay in repolarization of the terminal and an increase in action-potential duration. The increase in action-potential duration leads to an increase in Ca^{2+} influx during the action potential, leading to an increased vesicle fusion. [From Kandel ER et al. (1983) Classical conditioning and sensitization share aspects of the same molecular cascade in *Aplysia*. *Cold Spring Harbor Symp Quant Biol* 47: 821–830.]

protein. This phosphorylation is apparently specific for dopamine, since other neurotransmitters do not affect DARPP-32 phosphorylation. In its phosphorylated form, DARPP-32 is a potent inhibitor of protein phosphatase-1, which is a broad-spectrum enzyme that dephosphorylates a number of phosphoproteins. By deactivating protein phosphatase-1

DARPP-32 would prevent the dephosphorylation of other proteins, thus potentiating the action of other kinases. The significance of this cascade for neuronal function is not yet certain.

Calcium/Calmodulin-Mediated Processes

Less is known about unique receptor and second messenger-generating systems for the calcium/calmodulin-mediated processes than is true of the cAMP system. One way that the calcium second messenger cascade can be initiated is through the phosphatidyl inositol system. Calcium may also enter the cell through voltage-sensitive calcium channels in the membrane; for example, some of the calcium/calmodulin-mediated processes that are thought to occur in presynaptic terminals are mediated by the calcium that enters the terminal during the action potential. Also, the calcium that enters the postsynaptic cell as a result of the channel opening associated with the NMDA receptor may function as a second messenger.

Much of what is known about the functional role of calcium/calmodulin second messenger systems in neurons comes from studies of the substrates of the calcium/calmodulin-dependent kinase and from studies of the kinase itself. Indeed, the existence of some calcium/calmodulin-dependent cascades was first inferred as a result of the discovery of substrates that were phosphorylated by specific kinases in particular situations.

The experimental strategy that led to the discovery of several kinases and substrates is exemplified by the work of P. Greengard and his colleagues (see Sieghart et al, 1978). The strategy was to incubate various samples (synaptic membrane fractions for example) with radiolabeled adenosine triphosphate (ATP) (the donor for phosphate) in the presence or absence of the second messenger (cAMP for example). It was found that different subsets of proteins were phosphorylated in the presence of cAMP than were phosphorylated under baseline conditions (Fig. 4.12). This permitted an identification of substrate proteins whose phosphorylation was cAMP-dependent. For example, a particularly prominent substrate of cAMP-dependent protein kinase is protein I, later termed synapsin because of its localization to presynaptic terminals. The neuronal localization of these cAMP-mediated cascades could be demonstrated by selectively destroying the neurons in a brain region with neurotoxins. After destroying the neurons, cAMP-dependent phosphorylation of certain proteins was no longer observed.

In other studies it was found that a different subset of proteins was phosphorylated when calcium was added. These studies thus allowed an identification of the substrates of the different kinases. Interestingly, some of the substrate proteins are phosphorylated by both cAMP and calcium/calmodulin-dependent kinases. Identical experimental strategies have been used to identify the substrate proteins that are phosphorylated by other second messenger cascades.

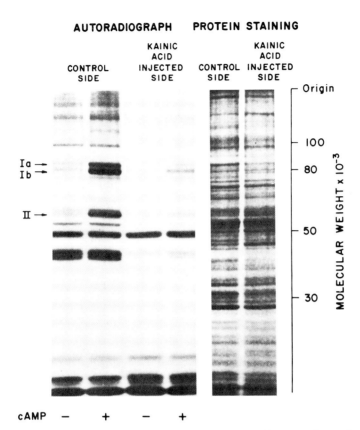

FIGURE 4.12. Cyclic AMP-dependent phosphorylation of neuronal proteins. Proteins from synaptic plasma membranes were isolated from rat caudate-putamen and incubated with labeled ATP in the presence or absense of cAMP. In the presence of cAMP there was a phosphorylation of three bands (Ia, Ib, and II). Destruction of neurons with kainic acid prior to isolation of synaptic plasma membranes eliminated the cAMP-stimulated phosphorylation. Thus, either the protein kinase, the substrate proteins, or both are eliminated when neurons are destroyed, suggesting that the cascade is present in neurons. [From Sieghart W (1978) Neuronal localization of specific brain phosphoproteins. *Brain Res* 156: 345–350.]

As noted above, one of the most prominent substrates proteins for both cAMP-dependent and calcium/calmodulin-dependent protein kinase is termed protein I or synapsin. This protein is neuron-specific, and within neurons it is found primarily in association with synaptic vesicles in terminals. The association between synapsin and synaptic vesicles suggested that synapsin might regulate vesicle function, thus providing a potential mechanism for presynaptic modulation of neurotransmission.

Electrophysiological studies of neurotransmitter release at the squid giant synapse have provided support for the hypothesis that phosphorylation of synapsin regulates neurotransmitter release. For example, injections of purified calcium/calmodulin-dependent protein kinase into the presynaptic terminal lead to a substantial release in neurotransmitter release without altering the calcium flux across the presynaptic terminal membrane (Fig. 4.13). It is thought that the injections of the kinase lead to an increased phosphorylation of synapsin, which then dissociates from the vesicles, and increases the probability of vesicle fusion during the calcium current that occurs during the presynaptic action potential.

Hints about the functional role of particular second messenger cascades can also be gained through studies of the localization of the kinases. For example, calcium/calmodulin-dependent protein kinase is highly concentrated in the postsynaptic membrane specialization of type I synapses on many CNS neurons. Indeed, this protein makes up a very large percentage

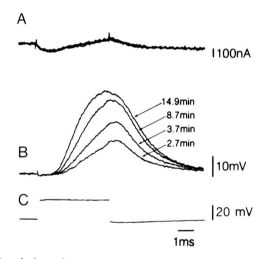

FIGURE 4.13. Regulation of neurotransmitter release by calmodulin kinase. Calmodulin kinase was injected directly into the presynaptic terminal of the squid giant synapse. The terminal was depolarized by a voltage step (bottom trace). The upper trace indicates the Ca^{2+} current evoked during the voltage step. The middle trace indicates the EPSP at various times after the injection of purified calmodulin kinase II. Note the progressive increase of EPSP amplitude after the injection of kinase. The injection of kinase is thought to result in an increase in the phosphorylation of calmodulin kinase substrates, including synapsin. The phosphorylation of synapsin leads to its dissociation from synaptic vesicles, leading to enhanced vesicle fusion during the presynaptic Ca^{2+} current. [From Llinas R et al. (1985) Intraterminal injection of synapsin I or calcium/calmodulin-dependent protein kinase II alters neurotransmitter release at the squid giant synapse. *Proc Natl Acad Sci USA* 82: 3035–3039.]

of the protein of the specialization. This is of considerable interest, since many of the synapses with the calcium/calmodulin-dependent kinase are also thought to possess NMDA receptors, which are thought to be calcium channels. Because of the localization of calmodulin and the kinase in the membrane specialization, any calcium that enters as a result of the activation of the NMDA receptor would have easy access to calmodulin and the calmodulin kinase. In turn, the kinase could phosphorylate any of a number of molecules involved in synaptic function, including receptors or channels.

Possible Role of NMDA Receptors in Synaptic Plasticity

As noted above, one receptor system that may use calcium as a second messenger is the NMDA receptor. The precise linkage between the receptor and calcium/calmodulin-dependent cellular processes has not yet been defined, but there is good evidence that the NMDA receptor does initiate a cascade that is important for synaptic modulation. This conclusion is based upon studies of the cellular mechanisms underlying a very long-lasting form of synaptic plasticity termed *long-term potentiation* (LTP).

Long-term potentiation of synaptic transmission occurs in certain CNS pathways after brief periods of high-frequency activation. The phenomenon was initially discovered in the hippocampus, where short periods of high-frequency stimulation of afferent pathways were found to induce increases in synaptic potency that persist for days and sometimes weeks (Fig. 4.14). The induction of LTP depends upon associative processes; in particular, activation of afferent pathways leads to LTP if and only if presynaptic activity occurs concurrently with a sufficient level of postsynaptic depolarization. There has been enormous interest in this form of synaptic plasticity because it is long lasting and associative. In fact, it is thought that cellular processes similar to those responsible for LTP may be the cellular mechanism for long-term information storage in the CNS.

The NMDA receptors have been implicated in the induction of LTP by several observations. First, LTP occurs in given afferent pathways only when the postsynaptic cell is sufficiently depolarized during the period of activation. Thus, the induction of LTP depends upon *voltage-dependent processes* similar to those required to activate the calcium channel associated with the NMDA receptor. Second, the induction of LTP depends upon calcium. Third, NMDA receptor antagonists effectively block the induction of LTP without substantially altering transmission via non-NMDA receptors. Thus, it is thought that when high-frequency activation of afferents occurs during periods of sufficient postsynaptic depolarization, calcium influx initiates some cellular process that leads to long-term modification of synaptic efficacy. The cellular nature of the changes induced by the calcium acting as a second messenger must still be defined.

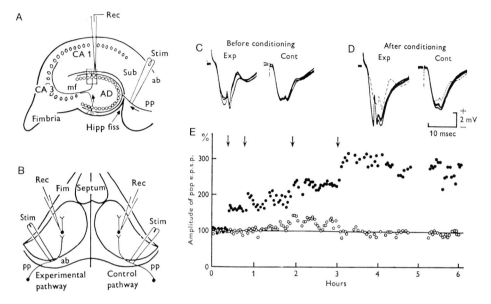

FIGURE 4.14. Long-term potentiation of synaptic efficacy in the hippocampal formation. *A* Illustration of a cross-section through the hippocampus of the rabbit showing the position of stimulating and recording electrodes when activating the perforant path (pp) from the entorhinal cortex to the area dentata (AD). This pathway terminates on the distal dendrites of the granule cells of the area dentata. *B* Dorsal view of the hippocampus illustrating the arrangement when the contralateral perforant pathway serves as a control for the ipsilateral pathway. *C* Responses evoked by experimental and control pathways prior to the induction of LTP. *D* Responses evoked by experimental and control pathways after conditioning stimulation of the experimental pathway. Note increases in response amplitude in the experimental pathway. Conditioning stimulation consisted of trains of stimuli (15/ s for 10 s) that were given at the *arrows*. *E* Time course of the increases in evoked-response amplitude over the experimental pathway (*filled circles*) in comparison with the control pathway (*open circles*). [From Bliss TVP, Lomo T (1973) Long-lasting potentiation of synaptic transmission in the dentate area of the anesthetized rabbit following stimulation of the perforant path. *J Physiol* 232: 331–356.]

Second Messengers and Cellular Compartmentation

Second messenger-linked receptors initiate a wide range of processes within neurons that lead to altered cellular function. These processes can operate at the stage of synaptic transmission, either by affecting presynaptic or postsynaptic processes, or can affect general membrane properties (i.e., potassium channels). Each second messenger is targeted to a specific protein kinase, which in turn has one or more specific substrates; the response to the second messenger depends upon the nature of the kinases and substrate proteins that are present in the cellular compartments that

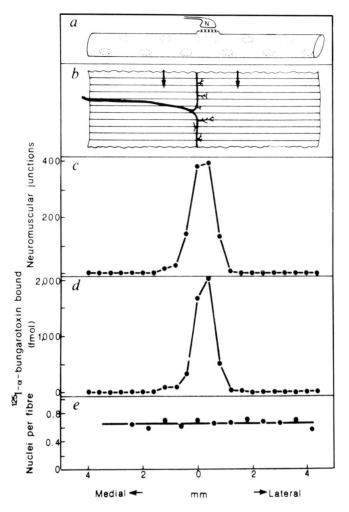

FIGURE 4.15. Localization of mRNA for ACh receptors (AChR) beneath synaptic sites in muscle. *A* Location of neuromuscular junctions in mouse diaphragm muscle. Muscle fibers in the diaphragm are multinucleate and receive a single contact from a motor nerve (N). ACh receptors (x) are concentrated in the postsynaptic membrane beneath the contact. *B* A strip of the diaphragm illustrating the alignment of adjacent muscle fibers. The motor axons branch and form synapses on each muscle fiber near the midpoint of the fiber. *C* Distribution of ACh receptors in the same sections as assessed by binding of ^{125}I-α-bungarotoxin. Note that receptors are concentrated in the center of the muscle fibers at synaptic sites. *E* Distribution of nuclei along the muscle fiber. Note that nuclei are uniformly distributed along the length of the fiber. [Reprinted by permission from *Nature* 317: 66–68. Copyright © 1985 Macmillan Magazines Ltd.]

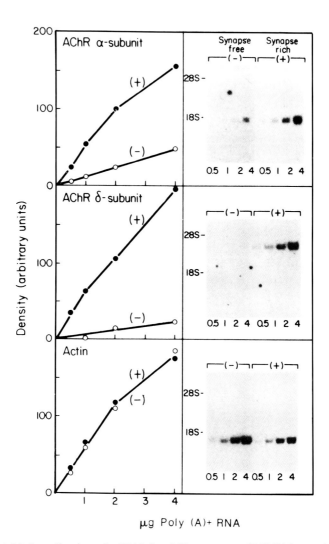

FIGURE 4.16. Localization of mRNA for ACh receptors (AChR) beneath synaptic sites in muscle. To determine the localization of mRNA for ACh receptors, diaphragms from mice were cut into synapse-rich and synapse-free strips (*arrows* in Fig. 4.15B). Poly (A)$^+$ RNA was purified from the pooled samples and fractionated on agarose gels. The mRNA was transferred to nitrocellulose and hybridized with a ^{32}P-labeled cDNA probe to mRNA for two subunits of the ACh receptor, and with a cDNA probe to the mRNA for actin. The autoradiographs on the right indicate the amount of mRNA present. Note that the mRNAs for both subunits of the ACh receptors are more concentrated in the samples from synapse-rich areas of the muscle fiber, whereas the mRNA for actin was present in both synapse-free and synapse-rich regions. [Reprinted by permission from *Nature* 317: 66–68. Copyright © 1985 Macmillan Magazines Ltd.]

have the elevated concentrations of the second messenger. Thus, specificity of action of second messenger depends upon *intracellular compartmentation*, that is, on the intracellular domain in which the second messenger is elevated and on the kinases and substrate proteins that are present there. By regulating the composition of these compartments, neurons can guarantee that appropriate cellular mechanisms are in place to carry out whatever modulations are necessary in response to neurotransmitters.

Synthesis and Turnover of Receptors

Relatively little is currently known about how receptors are synthesized and delivered to synaptic sites in neurons, or about receptor turnover. Some information is available for the ACh receptor of muscle, however.

There is evidence to suggest that the ACh receptor is delivered to the postsynaptic membrane specialization in *coated vesicles*. Presumably, these fuse with the plasma membrane and deposit their collection of receptors. Exactly where in the cell the subunits of the receptor are synthesized and glycosylated is not clear, but the synthesis and posttranslational processing presumably occurs in elements of the endoplasmic reticulum and Golgi apparatus.

Recent studies using molecular biological techniques indicate that in mature muscles, the ACh receptor may actually be synthesized in cytoplasmic domains immediately beneath synaptic sites. An elegant study by Merlie and Sanes (1985) took advantage of the unique arrangement of neuromuscular synapses in the rat diaphragm to study the intracellular distribution of the mRNA for the ACh receptor. Motor axons terminate in a narrow strip in the middle of the individual muscle fibers comprising the diaphragm (Fig. 4.15). Thus, it is relatively easy to dissect regions of the muscle that are synapse-rich and synapse-free. When the mRNA from synapse-rich and synapse-free portions of the muscle was probed with cDNA to different subunits of the ACh receptor, it was found that the mRNA for the receptor was highly concentrated in the portion of the muscle fiber on which the motor nerve terminates (Fig. 4.16). However, the mRNAs for nonsynaptic proteins (such as actin) were uniformly distributed throughout the postsynaptic cell's cytoplasm. The selective localization of mRNA beneath the synaptic site means that the synthesis and posttranslational processing of the receptor could be regulated locally.

The selective positioning of polyribosomes beneath synaptic sites on CNS neurons indicates that neurons also synthesize some proteins in cytoplasmic microdomains associated with individual postsynaptic sites. In the case of neurons, however, the identity of the molecules that are synthesized beneath synapses must still be determined.

Supplemental Reading

General

McGeer PL, Eccles JC, McGeer EG (1987) *Molecular Neurobiology of the Mammalian Brain.* New York, Plenum Press

Ionophore-Linked Receptors

Changeux J-P (1987) The acetylcholine receptor molecule: allosteric sites and the ion channel. *Trends Neurosci 6:247–251*
Foster AC, Fagg GE (1987) *Taking apart NMDA receptors. Nature* 329:395–396
Grenningloh G, Rienitz A, Schmitt B et al. (1987) The strychnine-binding subunit of the glycine receptor shows homology with nicotinic acetylcholine receptors. *Nature* 328:215–220
Schofield PR, Darlison MG, Fujita N et al. (1987) Sequence and functional expression of the $GABA_A$ receptor shows a ligand-gated receptor super-family. *Nature* 328:221–227

Second Messenger Cascades

Berridge MJ (1984) *Biochem J* 220:345–360
Greengard P (1979) Cyclic nucleotides, phosphorylated proteins, and the nervous system. *Fed Proc* 38:2208–2217
Hemmings HC Jr, Walaas SI, Ouimet CC et al. (1987) Dopaminergic regulation of protein phosphorylation in the striatum: DARPP-32. *Trends Neurosci* 10:377–382
Dunlap K, Holz GG, Rane SG (1987) G proteins as regulators of ion channel function. *Trends Neurosci* 10:244–247
Nishizuka Y (1984) Turnover of inositol phospholipids and signal transduction. *Science* 225:1365–1369
Nishizuka Y (1984) The role of protein kinase C in cell surface signal transduction and tumour promotion. *Nature* 308/19:693–697

Second Messengers and Functional Modifications

Brunelli M, Castellucci V, Kandel ER (1976) Synaptic facilitation and behavioral sensitization in *Aplysia:* possible role of serotonin and cyclic AMP. *Science* 194:1178–1181
Goelet P, Castellucci VF, Schacher S et al. (1986) *Nature* 322:419–422
Kandel ER, Schwartz JH (1982) Molecular biology of learning: modulation of transmitter release. *Science* 218:433–443
Nestler EJ, Greengard P (1983) Protein phosphorylation in the brain. *Nature* 305:583–589

Long-term Potentiation

Bliss TVP, Gardner-Medwin AR (1973) Long-lasting potentiation of synaptic transmission in the dentate area of the unanesthetized rabbit following stimulation of the perforant path. *J Physiol* 232:357–374

Bliss TVP, Lomo T (1973) Long-lasting potentiation of synaptic transmission in the dentate area of the anesthetized rabbit following stimulation of the perforant path. *J Physiol* 232:334–356

Smith SJ (1987) Progress on LTP at hippocampal synapses: a postsynaptic Ca^{2+} trigger for memory storage. *Trends Neurosci* 10:142–144

Cellular Mechanisms of Receptor Synthesis and Turnover

Bursztajn S, Fischbach GD (1984) Evidence that coated vesicles transport acetylcholine receptors to the surface membrance of chick myotubes. *J Cell Biol* 98:498–506

Devreotes PN, Fambrough DM (1976) Synthesis of acetylcholine receptors by cultured chick myotubes and denervated mouse extensor digitorum longus muscles. *Proc Natl Acad Sci USA* 73:161–164

Merlie JP, Sanes JR (1985) Concentration of acetylcholine receptor mRNA in synaptic regions of adult muscle fibers. *Nature* 317:66–67

CHAPTER 5

Histogenesis of the Nervous System

Introduction

Given the complexity and precision of CNS organization, the assembly of the elements of the nervous system must certainly be among the most remarkable feats accomplished during embryonic development. An understanding of the processes that shape the nervous system must begin with a description of how the resident cells are generated, and how they come to be positioned in the appropriate location, that is, upon the process of *proliferation* of the cells of the nervous system and *migration* of these cells from the site where they are generated to their final residence. With these processes comes the *determination* of how the cells will differentiate. These topics represent the core of the study of nervous system *histogenesis*. Finally, the key to nervous system function lies in the specific connections that are formed between neurons. This chapter considers the topic of histogenesis, while Chapters 6 and 7 consider how connections form between the elements of the nervous system.

In studying nervous system development, some themes from classical embryology should be recalled. In general, cellular differentiation involves an interaction between *intrinsic influences* that arise from the commitment of a cell to a program of differentiation that is played out with little or no modification by subsequent events and *epigenetic influences* that are extrinsic to the cell. In turn, epigenetic influences can take different forms. For example, an extrinsic signal can initiate a program of differentiation within a cell that will progress independently; this represents the phenomenon of *induction*. Extrinsic signals can also operate more or less continuously to shape and mold cellular differentiation; in this case it is a series of extrinsic influences rather than a single triggering event that determines the pattern of differentiation.

An excellent example of an inductive phenomenon that is important for nervous system development involves the induction of the neural tube. Hans Spemann, in work that eventually led to the Nobel Prize in 1935, found that the formation of the neural plate and tube was induced by tissue present in the dorsal lip of the blastopore (Spemann 1938). This conclusion was based on the fact that portions of the dorsal lip could induce the formation of a neural plate in unusual locations when transplanted from one gastrula to another. Subsequent experiments indicated that it was the mesoderm, including the notochord, that was the crucial region for this form of induction. Experiments that involved transplanting pigmented cells into unpigmented hosts revealed that donor and host cells were mixed in the reorganized tissue, although the mesoderm and notochord usually arose from the transplant whereas the nervous system arose from the host. These observations suggested that certain cells in the transplant served as an *organizer* for future differentiation of the tissue around them.

Spemann's experiments on the organizer clearly demonstrated an epigenetic influence of one type of tissue on the development of the nervous system. Experiments that seek to reveal epigenetic influences are generally of two forms. The first, exemplified by the organizer experiment, demonstrates the induction of a pattern of differentiation where it otherwise would not occur. The second class of experiment involves the disruption of normal tissue interactions, asking whether such disruption interferes with normal development. The latter type of experiment is particularly prone to false-positive results. For example, some manipulations of the cellular environment may exceed the range of conditions to which cells can respond using the mechanisms that are normally at work. In this case the response may be a pathological one that is qualitatively different from that which operates during normal development.

When thinking about extrinsic or epigenetic versus intrinsic influences, it is important to recall that the commitment of the cell to a program of differentiation can and most likely does arise from earlier *extrinsic* influences (cell-cell interactions). Once initiated, such a program of differentiation is determined by *intrinsic* factors, but the initiation of the program depends upon an *extrinsic* event. Thus, the terms "intrinsic" and "extrinsic" are useful in theory, and perhaps applicable at particular times, but are often not entirely separable. Neural development depends upon a continuous interaction between factors intrinsic to cells (determined by programmed differentiation set into motion early in the lifetime of the cell) and extrinsic factors arising from cell-cell and cell-environment interactions. The key to understanding the mechanisms of differentiation is understanding these *interactions* that occur between intrinsic and extrinsic influences.

The Formation of the Nervous System

Elements of the nervous system arise from three sources: (1) the *neural tube*, (2) the *neural crest*, and (3) *ectodermal placodes*. The neural tube gives rise to all of the nervous system except for the peripheral nervous

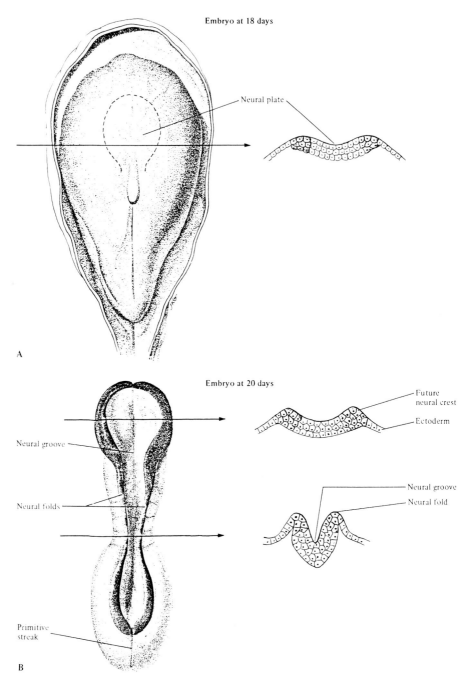

Embryo at 18 days

Neural plate

A

Embryo at 20 days

Future
neural crest

Ectoderm

Neural groove

Neural groove

Neural fold

Neural folds

Primitive
streak

B

FIGURE 5.1. Differentiation of the neural plate into the neural tube (human embryo: 18, 20, 21, and 24 days). The neural plate folds to form a groove and then a tube. The cells of the neural crest separate from the neural plate at the time of closure. [From Heimer L (1983) *The Human Brain and Spinal Cord*. New York, Springer-Verlag.]

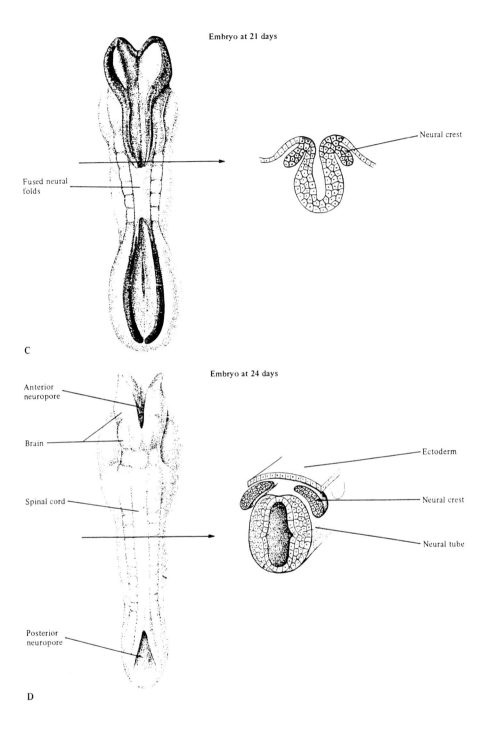

Embryo at 21 days

Neural crest

Fused neural
folds

C

Embryo at 24 days

Anterior
neuropore

Brain

Spinal cord

Ectoderm

Neural crest

Neural tube

Posterior
neuropore

D

system and the sensory epithelia. The neural crest gives rise to spinal and autonomic ganglia (the sympathetic and parasympathetic nervous systems), to the glial cells of the peripheral nervous system, and to nonneural tissue, including pigment-producing cells (melanocytes), chromaffin cells of the adrenal medulla, cells of the pia mater and arachnoid mater, and portions of the facial skeleton, including parts of the ossicles of the middle ear. The ectodermal placodes arise from nonneural ectoderm, and among other tissues, give rise to the sensory epithelia and to some of the ganglion cells in the ganglia of the cranial nerves.

The Formation of the Neural Tube: Neurulation

The *neuroectoderm* develops along the dorsal midline of the embryo. When the neuroectoderm first becomes apparent, it appears as a slight depression along the dorsal midline. Transverse sections reveal that the area is slightly thickened—compared with the *somatic ectoderm* lying more laterally— as a result of elongation of the cells (Fig. 5.1). Between the neuroectoderm and the somatic ectoderm is the *neurosomatic junctional region.* The neuroectoderm forms the *neural plate* (or *medullary plate*), which subsequently gives rise to most of the nervous system. The neurosomatic junctional region gives rise to the *neural crest.*

As development proceeds, the neuroectoderm continues to thicken, and the lateral edges of the neural plate begin to elevate and infold so as to form a *neural groove* bounded laterally by *neural folds.* The neural folds come together at the midline and fuse so as to form the *neural tube.* Closure of the neural tube begins in the upper cervical region and proceeds from anterior to posterior. As the neural tube forms, the cells of the neurosomatic junctional region separate from the neuroectoderm and somatic ectoderm and assume a position dorsolateral to the neural tube forming the *neural crest* (Fig. 5.1). For a time, the ends of the neural tube remain open, leaving the *anterior* and *posterior neuropores.* These subsequently close, leaving the lumen of the neural tube isolated from the rest of the embryo. The rostral end of the neural tube then gives rise to three vesicular swellings termed the *prosencephalic, mesencephalic,* and *rhombencephalic vesicles;* these give rise to the forebrain, midbrain, and hindbrain, respectively, while the caudal portion of the neural tube gives rise to the spinal cord. In humans the neuroectoderm forms at about 18 days of gestation, the process of neural tube formation begins toward the end of the third embryonic week, and the neuropores close during the fourth embryonic week.

Neural Crest Derivatives

The various cells that derive from the neural crest reach their final resting place after migrating from the area dorsolateral to the neural tube. As noted above, the neural crest cells give rise to a number of different cell types, including the neurons of the sympathetic and parasympathetic nervous system. How cells become determined to differentiate in a particular way is not entirely clear, but it does seem that this determination is made very early on in development. In the case of the neurons, however, there are at least some epigenetic influences that can be identified.

There is good evidence for epigenetic influences in the determination of the transmitter type used by the neurons of the sympathetic and parasympathetic nervous system. In general, most sympathetic ganglion cells are adrenergic, whereas parasympathetic ganglion cells are cholinergic. Both types originate from the neural crest, and both are of a neuronal lineage; the two differ in terms of their route of migration and their final resting place. Thus, the question arose whether transmitter choice is determined before migration or by the route of migration or the local tissue environment in which the neuron comes to rest.

Experiments to evaluate the role of the local tissue environment in differentiation of neural crest derivatives were carried out by N. LeDouarin and M.-A. Teillet (1974); these studies took advantage of the fact that neural crest cells from chicks and quails can be distinguished on the basis of differences in nucleolar size. In chicks and quail the cholinergic parasympathetic ganglia arise from the so-called vagal region (the first seven somites), whereas the adrenergic neurons of the sympathetic chain arise from the portion of the neural crest in the trunk. Each neuron type thus arises in a different portion of the neural crest and migrates via a different route. When neural crest cells from the trunk were transplanted into the vagal region, the cells migrated as would "native" vagal neural crest cells and gave rise to cholinergic neurons. Conversely, when cells from the vagal region of the neural crest were transplanted to the trunk region, they migrated as would native trunk crest cells and gave rise to adrenergic ganglion cells of the sympathetic nervous system.

Subsequent experiments have revealed some of the factors that can affect the choice of neurotransmitter by neural crest derivatives. For example, when sympathetic neurons obtained from late embryos are grown for several weeks in culture, the transmitter that is expressed by these cells is determined by the culture conditions. Also, if sympathetic neurons are grown in medium obtained from cultures of heart cells (heart-conditioned medium), the sympathetic neurons lose their adrenergic properties and begin to synthesize ACh (Fig. 5.2). In an elegant series of studies, E.J. Furschpan et al. (1975) grew small numbers of ganglion cells (sometimes single neurons) on islands of heart cells. Using electrophysiological techniques, they were able to document that under the influence of heart cells, individual neurons switched from adrenergic to cholinergic transmission, passing through a stage in which both neurotransmitters were released. These studies strongly suggest that the differentiation of neurons of neural crest origin depends on the tissue environment in which the cells reside after the completion of migration.

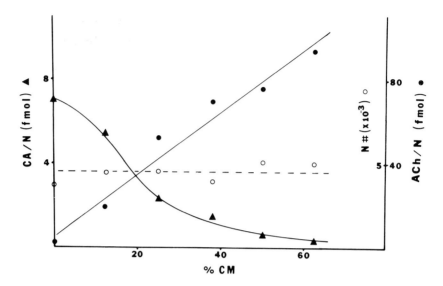

FIGURE 5.2. Determination of neurotransmitter type in neurons of neural crest origin. Cells of the neural crest give rise to neurons of both parasympathetic and sympathetic nervous systems. The transmitter in postganglionic neurons of the parasympathetic nervous system is ACh, whereas norepinephrine is the transmitter in the postganglionic neurons of the sympathetic nervous system. In culture, the type of transmitter expressed by neurons of neural crest origin depends on the culture conditions. The addition of various concentrations of "conditioned medium" (CM) taken from flasks of heart cells decreases the production of catecholamines (CA) and increases the production of ACh. Thus, under the influence of heart-conditioned medium, the neurons express a parasympathetic phenotype. [From Patterson PH, Chun LLY (1977) The induction of acetylcholine synthesis in primary cultures of dissociated rat sympathetic neurons. *Dev Biol* 56: 263–280.]

The Ectodermal Placodes

The ectodermal placodes that are especially important for the development of the nervous system are the *nasal placode* and the *acousticovestibular placode*. The nasal placode gives rise to the olfactory epithelium, including the olfactory receptors. The receptor neurons differentiate into bipolar neurons with a peripheral process that is specialized to detect airborne molecules that contact the overlying mucous layer and a central axonal process that projects into the olfactory bulb. The acousticovestibular (otic) placode gives rise to the sensory epithelia of the inner ear and to the ganglion cells of the acoustic or spiral ganglion and the vestibular ganglion. These ganglion cells also eventually become bipolar, with a peripheral process that terminates upon hair cells and a central process that projects into the brainstem. In addition, some of the ganglion cells of the trigeminal, glossopharyngeal, and vagal ganglia arise from the placodes.

Proliferation and Migration of Neurons and Glia

Histogenesis in the Neural Tube

With the formation of the neural tube, the columnar epithelium of the wall of the neural tube is converted into a *pseudostratified* epithelium in which mitotic figures are found exclusively at the luminal surface. Conversely, the cells that are not actively dividing extend the full width of the *epithelium*, having one cytoplasmic process in contact with the *basement membrane (basal lamina)* and the other process facing the lumen of the neural tube (Fig. 5.3). This converted epithelium is called the *germinal epithelium, neuroepithelium, germinal matrix,* or *ventricular layer*. The last term is especially common when referring to the layer at later stages of development, after the formation of mantle and marginal layers (see below).

The Cell Cycle in the Neuroepithelium

Studies using 3H-thymidine as a tracer have helped to reveal the sequence of events during proliferation. Systemic injections of 3H-thymidine result in the labeling of cells that are actively synthesizing DNA at the time of

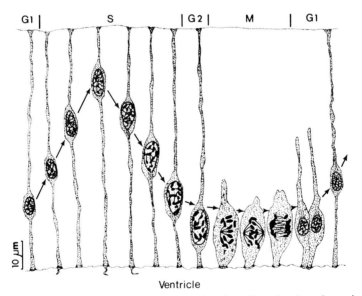

Ventricle

FIGURE 5.3. The cell cycle in neural tube. During the cell cycle, there is a migration of the nucleus of the germinal neuroepithelial cell. Before mitosis, during the late S phase, the nucleus migrates toward the ventricle. The peripheral process of the cell is withdrawn and cytokinesis occurs. The distal process is again extended, and the cycle begins again. [From Jacobson M (1978) *Developmental Neurobiology* (ed 2). New York, Plenum Press. After Sauer FC (1935) *J Comp Neurol* 62: 377–405.]

the injection. The labeled precursor is rapidly excreted, so that the labeled precursor is only available for incorporation for one to two hours. If animals are killed soon after the injection and prepared for autoradiography, only cells that were actively synthesizing DNA during the "pulse" of the 3H-precursor are labeled. However, if the animals are allowed to survive for varying periods after the pulse, the labeled cells may continue to divide, diluting the label by one half at each division. By varying the survival interval, it is possible to identify cells that underwent their terminal mitosis during the pulse, as well as cells that have undergone one or more subsequent divisions after the initial labeling.

In the case of cells in the neuroepithelium, when embryos are prepared for autoradiography one to two hours after labeling with 3H-thymidine, labeled nuclei are found primarily in the external portion of the neuroepithelium (Fig. 5.4A). Thus, DNA synthesis occurs in cells when they are located away from the ventricular surface. At longer intervals (six hours) labeled nuclei are concentrated at the ventricular surface, and many of the labeled neurons are undergoing mitosis. At still longer intervals (12 and 48 hours, see Fig. 5.4), the extent of labeling is less because the label has been diluted by half at the time of cell division. In addition, the labeled nuclei are again found in the external portion of the ependymal zone. These results suggested the sequence of events depicted in Figure 5.3. As cells in the neuroepithelium prepare to divide, the cells in the S phase of the cycle withdraw the process that is apposed to the basement membrane, their nuclei migrate toward the luminal surface, and the cells round up. After mitosis, the daughter cells again extend a process to contact the basement membrane, and their nuclei migrate to a basal position. This cycle is then repeated until the cells leave the proliferative phase and begin their migration.

3H-thymidine autoradiography can also determine the duration of the cell cycle. During the time that 3H-thymidine is available for incorporation, the cells that are synthesizing DNA prior to mitosis incorporate the precursor. Thus, at some time after the pulse (determined by the average time between DNA synthesis and the initiation of mitosis), essentially all mitotic figures are labeled. As this population of cells enters the G1 phase (presynthetic gap), the percentage of labeled mitotic figures decreases to a very low level (these cells had been synthesizing DNA after 3H-thymidine was no longer available). Subsequently, when the cells that had been labeled during the pulse again begin to divide, there is a second peak in the percentage of labeled mitotic figures (Fig. 5.5). In this way the duration of the various phases of the cell cycle can be determined. In general, cell cycle times tend to be shorter at early stages of development and to increase at later stages.

Prior to the recognition that the neuroepithelium consisted of a single cell type at different stages of the cell cycle, early embryologists (particularly W. His) thought that neurons arose from the dividing *germinal cells*

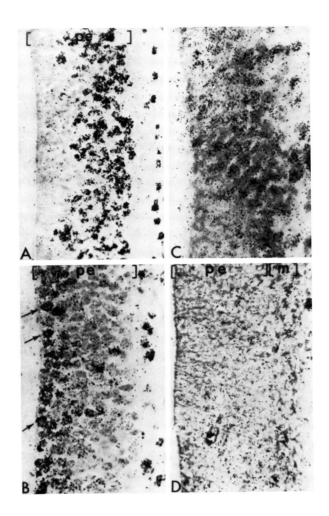

FIGURE 5.4. Use of 3H-thymidine autoradiography to label dividing neurons reveals the migration of the nuclei of neuroepithelial cells during the cell cycle. 3H-thymidine was injected intravenously into pregnant mice on the 11th day of gestation. Autoradiographs were prepared of sections taken through the cerebral vesicles. Cells synthesizing DNA at the time of the injection in preparation for mitosis are heavily labeled. In cells that continue to divide, the label is sequentially diluted by ½ at each division. *Brackets* indicate the ependymal zone (pe) and the mantle zone (m). *Arrows* indicate mitotic figures. *A* One hour after the 3H-thymidine injection; *B* six hours after 3H-thymidine; *C* 24 hours after 3H-thymidine; *D* 48 hours after 3H-thymidine. The wall of the cerebral vesicle is about twice as thick in the 13-day-old embryo (D) as in the 11-day embryo (*A*); thus, the photographic magnification is reduced. *A-C* 400X; *D*, 225X. [From Sidman RL, Miale IL, Feder N (1959) Cell proliferation and migration in the primitive ependymal zone; an autoradiographic study of histogenesis in the nervous system. *Exp Neurol* 1: 322–333.]

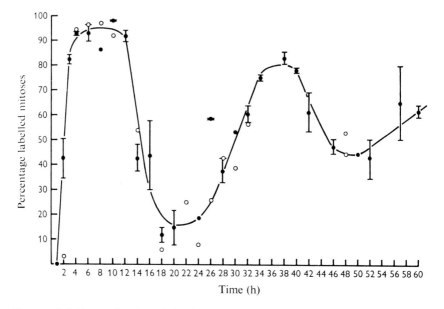

FIGURE 5.5. Determination of the duration of the cell cycle. The figure illustrates the percentage of labeled mitotic figures in the retina at various times after an injection of 3H-thymidine. Immediately after the injection, all mitotic figures are labeled, since DNA synthesis was occurring in these cells prior to mitosis. The percentage of labeled mitotic figures drops and then increases as the cells that were dividing during the pulse of 3H-thymidine again undergo mitosis. The time between peaks is the duration of the cell cycle. [From Denham S (1967) A cell proliferation study of the neural retina in the two-day rat. *J Embryol Exp Morphol* 18: 53–66 with permission of Company of Biologists Ltd.]

and that glial cells arose from nonneuronal epithelial cells that were termed *spongioblasts*. This view is now recognized as being in error, but the term spongioblast is still sometimes used. In particular, some tumors of the developing nervous system are termed *spongioblastomas*.

The Formation of the Ventricular, Mantle, and Marginal Zones

At some point, cells undergo a *terminal mitosis* and cease further synthesis of DNA, becoming arrested in the G1 phase of the cell cycle. Postmitotic cells then begin to migrate toward the basal lamina and past the nuclei of the neuroepithelial cells (Fig. 5.6). As the cells accumulate in this zone, they form the *mantle layer*, which becomes thicker as a result of the addition of more and more postmitotic cells. Between the mantle layer and the basal lamina is the *marginal zone*, which initially contains the basal processes of the neuroepithelial cells, but is then invaded by the processes of the cells in the mantle layer and by ingrowing processes from other parts of the nervous system. As the neurons in the mantle layer begin to

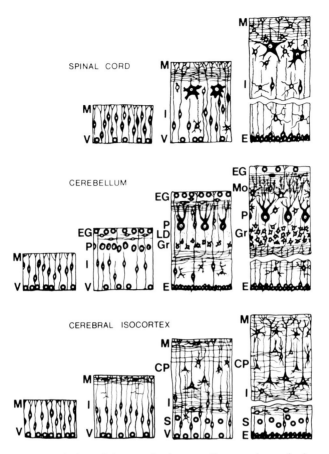

FIGURE 5.6. Differentiation of the mantle, intermediate, and marginal zones in the spinal cord, cerebellum, and cerebral cortex. Progressively later stages of development are shown from left to right. M = marginal zone; I = intermediate zone; E = ependymal layer; V = ventricular zone; P = Purkinje cell layer; EG = external granule layer; Gr = granule layer; Mo = molecular layer; CP = cortical plate; S = subventricular zone or subependymal zone. Stars = mitotic figures. [From Jacobson M (1978) *Developmental Neurobiology* (ed 2). New York, Plenum Press.]

differentiate, elaborating their axons and dendrites, an *intermediate layer* is formed; this increases in thickness as the neural tube differentiates into the spinal cord (Fig. 5.6).

Commitment of Cells to Neural and Glial Lineages

It is not known when cells become committed to the neural and glial lineages. Indeed, it is not even known whether both daughter cells of the last mitosis prior to migration have the same fate, although many inves-

tigators favor the view that different precursor populations exist for neurons and glia. When the cells are in the mantle layer, there is no morphological distinction that can be drawn between *glioblasts* and young
neurons. However, once migration has begun, the future neurons are no
longer capable of cell division, whereas glial cells are; indeed, glial cells
retain the capacity for further division throughout the life of the organism.
This suggests that cells are fully committed to differentiate into neurons
or glia at the commencement of migration. Because young neurons are
not capable of further cell division, the term "young neuron" is more
appropriate than the term *neuroblast*, which is sometimes used, whereas
glioblast is appropriate since these cells are still capable of cell division.

Although cell types cannot be distinguished morphologically until they
begin to differentiate, recent studies using molecular markers also suggest
an early commitment of cells to one or the other lineage. For example,
an accepted molecular marker for glial cells is glial fibrillary acidic protein
(GFAP, see Chapter 2). Using immunocytochemical techniques, it has
been found that some cells in the ventricular zone express GFAP. Assuming that neurons do not transiently express GFAP, this evidence suggests that glial cells are already determined in the ventricular zone before
migration. The flaw with such studies is that it is entirely possible that
cells *do* express proteins of a different cell type early in their differentiation.
Thus, the issue of when a cell becomes committed to differentiate along
neuronal or glial lines is still open.

The basic pattern of proliferation, migration, and differentiation that is
seen in the neural tube is embellished during the development of other
parts of the nervous system. In general, laminated structures (cerebral
cortex, archicortex, and cerebellum) exhibit a similar basic histogenic
pattern, with some unique histogenic variants in the different regions due
to the existence of secondary germinal zones. Nonlaminated structures
exhibit a slightly different pattern.

Histogenesis in Laminated Structures; the Cerebral Cortex

For the cerebral cortex, the histogenic pattern resembles that seen in the
spinal cord, except that four layers can be distinguished (Fig. 5.6). The
cells of the cortex arise from a *ventricular zone* analogous to the neuroepithelium in the neural tube. This zone appears first, gives rise to cells
that form the other layers, and then disappears at an early stage of development, differentiating into the *ependymal layer*. A *subventricular zone*
or *subependymal zone* also contains proliferating cells; cells in this zone
continue to proliferate after the ventricular zone has differentiated into
the ependymal layer. As the cells begin their outward migration, two additional layers are formed: an *intermediate zone* and a *marginal zone*.
These are analogous to the similarly named zones in the neural tube. The
intermediate zone contains the differentiating neurons and glia and even-

tually the ingrowing axons. As the cells in the intermediate zone begin to differentiate, the layer of differentiating cells is termed the *cortical plate*.

An important feature of cortical histogenesis is that the neurons exhibit an *inside-out* gradient of histogenesis—that is, cells that undergo their final mitosis early in development are found in deep cortical layers, whereas later-forming cells are found in progressively more superficial layers. This pattern is also termed "stacking" because late-forming cells stack on top of neurons that have formed earlier.

Spatiotemporal gradients of histogenesis have been characterized by determining the "birthday" of cells at different locations in the mature nervous system using 3H-thymidine autoradiography. As noted above,

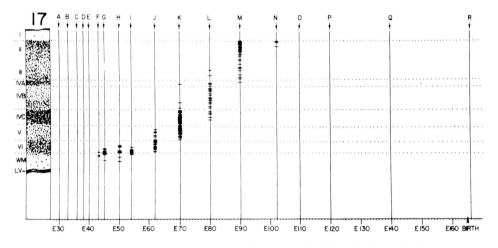

FIGURE 5.7. Inside-out gradient of cortical histogenesis. The "birthdays" of neurons in the cerebral cortex of the monkey were determined using 3H-thymidine autoradiography. 3H-thymidine is injected at the ages indicated by vertical lines A–R, and animals were allowed to develop to maturity. Cells undergoing their final mitosis at the time of the injection are heavily labeled. The figure illustrates the positions of heavily labeled neurons in the primary visual cortex (area 17) of juvenile monkeys that had been injected once with 3H-thymidine on the indicated embryonic days. The drawing on the left indicates cortical layers. Embryonic days are indicated on the horizontal line, starting at the left with the end of the first fetal month (E28) and ending at birth (E165). The relative position of labeled cells within the cortex is marked by the short horizontal bars on the vertical line. The total number of short horizontal bars on each line (except for N) indicates the number of labeled neurons encountered in one 2.5-mm strip of cortex. Since the number of labeled neurons decreases toward the end of histogenesis, a total of 80 strips of cortex were evaluated for N. [From Rakic P (1978) Neuronal migration and contact guidance in the primate telencephalon. *Postgrad Med J* 54: 25–40. After Rakic P (1974) Neurons in rhesus monkey visual cortex: systematic relation between time of origin and eventual disposition. *Science* 183: 425–427. Copyright 1974 by the AAAS.]

pulse labeling with 3H-thymidine results in the labeling of all cells that are synthesizing DNA prior to mitosis. If the 3H-thymidine is incorporated during the final mitosis, the label is retained throughout the life of the organism. If the cells are still dividing, however, the label is diluted by one half at each cell division. Thus, injections of 3H-thymidine at particular stages of development lead to a persistent labeling of the cells undergoing their final mitosis on that day. These labeled cells can be detected later in life using autoradiography (Fig. 5.7).

The existence of the inside-out gradient means that later-forming cells migrate past neurons that are already in place. This fact may have important implications for understanding the subsequent development of intracortical connections. Intracortical circuitry is characterized by a columnar organization such that cells in different cortical layers communicate with one another through interlaminar connections. The determination of intracortical connectivity may be partly due to interactions between neurons during migration.

Histogenesis in Cerebellar Cortex

The cerebellar cortex exhibits a histogenetic pattern that is a variant of that found in the cerebral cortex. The early differentiation of the cerebellar cortex proceeds like that of the cortex. A germinal layer in the roof of

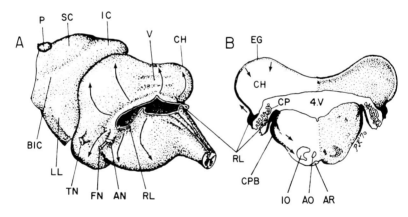

FIGURE 5.8. Migration of neurons from the rhombic lip to the external granule layer of the cerebellum and various brain stem nuclei. *A* Model of the human rhombencephalon at the end of the third month of gestation. *B* Cross-section through the rhombencephalon. RL = rhombic lip. *Arrows* indicate the route of migration of neurons. AO = accessory olive; AR = arcuate nucleus; BIC = brachium of the inferior colliculus; CH = cerebellar hemisphere; CP = choroid plexus; CPB = corpus pontobulbare; EG = external granule layer; FN = facial nerve; IC = inferior colliculus; IO = inferior olive; LL = lateral lemniscus; P = pineal gland; SC = superior colliculus; TN = trigeminal nucleus; V = vermis of the cerebellum; 4V = fourth ventricle. [From Sidman RL, Rakic P (1973) Neuronal migration, with special reference to developing human brain; a review. *Brain Res* 62: 1–35.]

the fourth ventricle (the ventricular zone) gives rise to the principal neurons (Purkinje cells), some of the large interneurons of the cerebellar cortex, and some glial cells. These cells migrate into an intermediate zone and begin to differentiate, forming a *cerebellar plate* similar to the cortical plate (Fig. 5.6). After the formation of the cerebellar plate, there is a migration of a population of true neuroblasts (i.e., cells still capable of mitosis) from the rhombic lip, which is a germinal zone in the wall of the fourth ventricle (Fig. 5.8). The cells migrate from the rhombic lip around the surface of the cerebellum to the zone immediately beneath the pia to form a second germinal zone *(the external granule layer)*. The cells in the external granule layer continue to divide for some time during cerebellar histogenesis. Indeed, the granule cells of the cerebellum—along with those of the hippocampus, which arise similarly (see below)—are the latest-forming neurons in the vertebrate brain. In humans the external granule layer does not disappear until about 600 days after birth. The most common

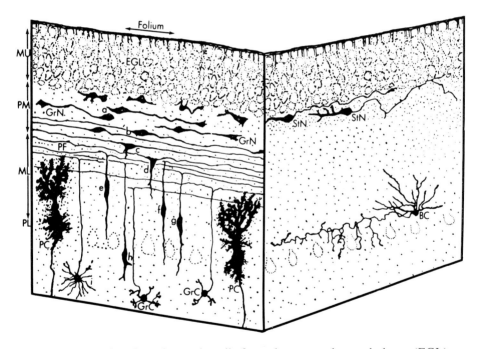

FIGURE 5.9. Migration of granule cells from the external granule layer (EGL) through the molecular layer (ML) of the cerebellum to their final destination in the internal granule layer. As the granule cells pass through the molecular layer, they give rise to their axonal processes, the parallel fibers (PF), see cell b. They then emit a single descending process (cell c and d), and the cell body migrates along this process into the internal granule layer (f–h). As this is occurring, the Purkinje cell (PC) elaborates its dendritic tree. [From Eccles JC (1970) Neurogenesis and morphogenesis in the cerebellar cortex. *Proc Natl Acad Sci USA* 66: 294–301.]

tumor of the CNS in children (medulloblastoma) appears to arise from portions of the external granule cell layer that persist abnormally into later life.

After undergoing their terminal mitosis, the cells in the external granule layer migrate down through the developing cerebellar cortex past the layer of Purkinje cell bodies to their final destination in the internal granule layer. As they pass through the molecular layer, they give rise to axonal processes from opposite poles of the cell body that develop into parallel fibers. A cytoplasmic process is then extended toward the Purkinje cell layer, and the cell body of the granule neuron migrates along this process into the internal granule layer (Fig. 5.9). The external granule layer also

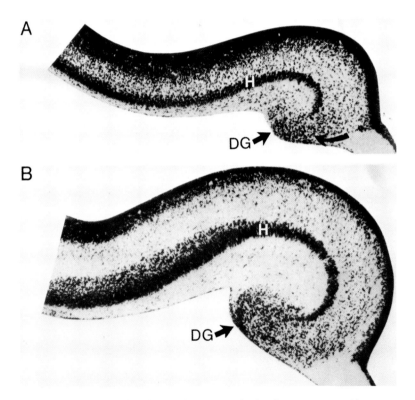

FIGURE 5.10. Histogenesis of granule neurons in the dentate gyrus. The true neuroblasts that are the precursors of granule cells of the dentate gyrus migrate from a small region in the ventricular zone into the hippocampus. Their route of migration is indicated by the *arrows* in *A*. After arriving in the substance of the hippocampus, the neuroblasts proliferate, forming the tightly packed layer of granule neurons. H = hippocampus; DG = dentate gyrus; *arrows* indicate the layer of dentate granule neurons. [After Stensaas L (1967a–d); *J Comp Neurol* 130: 149–162; *J Comp Neurol* 131: 409–422; *J Comp Neurol* 131: 423–436; *J Comp Neurol* 132: 93–108.]

gives rise to stellate and basket cells that migrate only as far as the molecular layer of the cerebellum. It has been proposed that further migration of the stellate and basket cells is prevented because these cells develop processes at right angles to the parallel fibers.

Histogenesis of the Hippocampal Formation

Another histogenetic variant is found in the hippocampal formation. The hippocampus develops from a portion of the neuroepithelium on the medial wall of the cerebral vesicle. The principal neurons of the hippocampus (the pyramidal cells) arise from the ventricular germinal zone, migrate

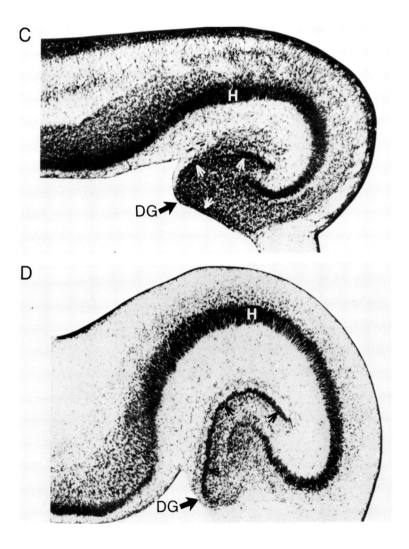

into an intermediate zone, and begin to differentiate. As the cells accumulate and begin to differentiate in the *hippocampal cortical plate,* the hippocampal formation infolds into the ventricle, but remains in contact with the portion of the neuroepithelium adjacent to the zone that gives rise to the choroid plexus of the lateral ventricle. As the hippocampal , cortical plate continues to differentiate, there is a secondary migration of true neuroblasts from the portion of the neuroepithelium that forms the wall of the lateral ventricle. These cells accumulate around one end of the layer that represents the hippocampal cortical plate and proliferate in neurons comprising the *dentate gyrus* (Fig. 5.10). The generation of the granule cells continues after birth. Indeed, there is evidence that granule cells continue to proliferate in adult animals.

Histogenesis in Nonlaminated Structures

The main difference in the histogenic pattern in nonlaminated (nuclear) portions of the brain is that the nuclei are formed as a result of an "outside-in" gradient of cell packing, that is, later-forming cells do not migrate past earlier-forming cells. This pattern is also termed "packing" in contrast to the "stacking" that is observed in cortical areas. The precise route of migration of postmitotic cells from the ventricular zone varies for different nuclei. In addition, some nuclei also arise as the result of migration of cells from one of the secondary germinal zones. Thus, for example, some neurons of the various brainstem nuclei arise from the secondary germinal zone in the rhombic lip (Fig. 5.8).

The conclusions that can be drawn regarding the genesis of neurons and glia are the following:

1. In any particular region, glial proliferation generally occurs at a later stage of development than neuronal proliferation.
2. Glial cells continue to be capable of mitosis throughout life, whereas virtually all neurons are incapable of cell division after the early postnatal period.
3. Within a given brain region, larger neurons tend to be born before smaller neurons.
4. When comparing sensory and motor nuclei at the same level of the neuraxis, neurons in the motor nuclei are born earlier.
5. Neurons in phylogenetically older regions of the brain are born earlier than neurons in younger regions.
6. Cortical regions exhibit an inside-out gradient of histogenesis (stacking), whereas nuclear regions exhibit an outside-in gradient (packing).

Migration of Neurons

In assembling the nervous system, there appears to be precise control over the routes taken by migrating neurons. The nature of this control is not known in most cases. Where neurons migrate over long distances, guidance may depend on physical cues in the tissue environment (i.e., differential substrate adhesion) or perhaps on chemotaxis.

What is known is that in the development of laminated structures, migrating neurons appear to follow the elongated processes of *radial glial cells*. These processes, termed radial fibers, extend from the ventricular

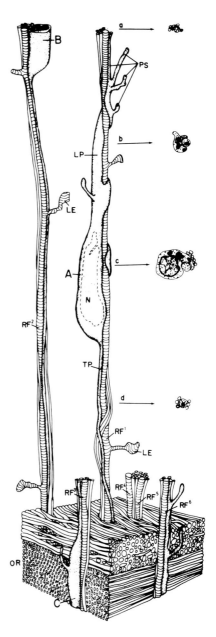

FIGURE 5.11. Relationship between migrating neurons in the cerebral cortex and radial glia. The figure illustrates a three-dimensional reconstruction based on electron micrographs of serial sections through the intermediate zone of the neocortex in a fetal monkey. The lower portion of the figure contains the uniformly oriented parallel fibers of the optic radiation (OR), with other less regularly oriented fibers above the OR. RF = radial fibers of the glia. Three migrating neurons that are closely apposed to the radial fibers are illustrated (A, B, C). The relationship between neuron A and the radial fiber of the glial cell is indicated in cross-section on the right-hand portion of the figure (a–d). LP = leading process of neuron; [From Rakic P (1972) Mode of cell migration to the superficial layers of fetal monkey neocortex. *J Comp Neurol* 145, 61–84.]

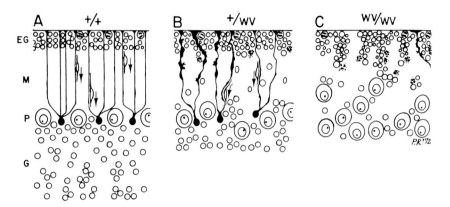

FIGURE 5.12. *A* Schematic illustration of the relationship between migrating granule neurons and Bergmann glia in the normal cerebellar cortex. Granule cells follow the radial processes of Bergmann glia in a manner similar to the migration of neurons in the cerebral cortex (Fig. 5.12). *B* and *C* illustrate the disruption of granule cell migration in the "weaver" mutation of mice. In heterozygous mice there is a reduction in the number of granule cells that successfully complete migration. In homozygous mice, almost no granule cells are successful in completing migration; instead, most of the granule cells degenerate at the border between the external granule layer and the molecular layer. [From Sidman RL, Rakic P (1973) Neuronal migration, with special reference to developing human brain: A review. *Brain Res* 62: 1–35.]

zone to the pial surface. Electron microscopic studies revealed that migrating neurons in the cerebral cortex are closely apposed to the radial glial processes during the course of their migration (Fig. 5.11). The same is true in the cerebellar cortex, except that the glial cells are more specialized *Bergmann glia* (see Chapter 2). In their migration from the external to the internal granule layer, the cerebellar granule cells appear to follow the processes of Bergmann glia (Fig. 5.12). In certain mutant strains of mice, the interaction between granule neurons and Bergmann glial processes is abnormal, and in such mutant mice, the migration of granule cells is severely disrupted. This evidence suggests that the relationship between migrating neurons and radial processes of glial cells is not simply fortuitous, but rather reflects an obligatory interaction.

Role of Cell Adhesion Molecules in Neural Histogenesis

An important process that occurs during development is the aggregation of like cell types, and the establishment of patterns through local interactions of cells. This is particularly true in the development of the nervous system, where like cells aggregate into characteristic nuclei or layers. Recent research suggests that *cell adhesion molecules (CAMs)* and *substrate adhesion molecules (SAMs)* play a major role in these aggregation processes, modulating cell-cell interactions during development. CAMs are thought to control the dynamics of morphogenesis because: (1) CAM expression follows a precise timetable during the establishment of borders of cell collectives, and (2) perturbation of CAM function dramatically alters pattern formation.

CAMs are glycoproteins that are expressed on cell surfaces that promote cell adhesion. Several types of CAMs have been identified, but the best characterized are *neural CAM (N-CAM), liver CAM (L-CAM), and neuron glia CAM (Ng-CAM)*. N-CAM and L-CAM appear very early in development, and both are expressed on all cells. After neural induction, the neural plate expresses only N-CAM, while the somatic ectoderm expresses primarily L-CAM. During secondary induction of endoderm and mesoderm, there are systematic transitions in the expression of N-CAM and L-CAM that seem to be important for the establishment of boundaries between cell collectives.

As the nervous system begins to differentiate, Ng-CAM is expressed on postmitotic CNS neurons. It is particularly prominent on growing neurites. Ng-CAM is also strongly expressed on the somata and leading processes of neurons that are migrating along radial glia. Both N-CAM and Ng-CAM are also present on Schwann cells and neurons in the peripheral nervous system, although N-CAM disappears from neural crest cells during migration, and reappears as the cells re-aggregate during ganglion formation.

During subsequent differentiation, N-CAM is expressed by all neurons. There is, however, a chemical conversion of N-CAM from an embryonic to an adult form that involves a loss of sialic acid residues, with a resulting change in binding properties. The expression of Ng-CAM diminishes with differentiation, particularly in myelinated fiber tracts.

Antibodies to the various CAMs have been found to disrupt morphogenesis in a manner consistent with the hypothesis that the CAMs play a role in pattern formation. For example, treatment with anti-N-CAM has been found to disrupt the formation of laminae in the retina. Anti-Ng-CAM was found to disrupt the migration of granule cells in the developing cerebellum, and to disrupt neurite fasciculation in culture. Clearly, much is still to be learned about the molecular mechanisms that lead to the formation of cell aggregates (nuclei) and cell laminae. However, the research on CAMs certainly represents a promising line of inquiry.

Programmed Cell Death in the Nervous System

The total number of neurons in any given brain region is remarkably constant from individual to individual; obviously, neuron number is closely regulated. Certainly part of the regulation is at the stage of proliferation and migration. The controls at these stages are not well understood, however. Another mechanism that regulates cell number during the development of many tissues is *programmed cell death*. This is also true in the nervous system, and at least some of the mechanisms that control this process of naturally occurring cell death are fairly well understood.

Some of the earliest work suggesting how cell number might be regulated involved the development of the spinal cord and sensory ganglia. The lateral motor column of the spinal cord consists of motor neurons that innervate peripheral musculature. The segments of the cord that innervate the limbs (brachial and lumbar segments) have many more motor neurons than segments of the cord that are not associated with the limbs. This relationship suggested to early investigators that cell number might somehow be regulated by the amount of available target.

Early studies amply supported this hypothesis. Thus, M.L. Shorey (1909) found that the removal of limb buds during early development resulted in a profound decrease in the number of motor neurons in the associated segment of the spinal cord. In addition, the size of sensory ganglia that would normally innervate the limb was decreased. Similarly, S. Detwiler (1920) demonstrated that if an extra limb was grafted onto the developing chick providing additional target tissue, there was an increase in the number of surviving cells in the spinal cord and an increase in the size of the sensory ganglia. These results have been confirmed and extended by several investigators, particularly V. Hamburger and his colleagues, so that the quantitative extent of the induced cell loss or cell sparing following manipulations of the available target tissue is now known (Fig. 5.13). What was not clear was how this regulation was accomplished. Initially it was thought that the target tissue might somehow regulate the extent of proliferation in the spinal cord. However, subsequent studies revealed that it was cell death that was regulated rather than cell proliferation.

Victor Hamburger, along with a number of his former students and colleagues, provided the most complete picture on the role that naturally occurring cell death plays in the development of the nervous system. Most

--▷

FIGURE 5.13. Reduction of naturally occurring cell death by increasing the available target tissue. *A* 6-day-old chick embryo with a supernumerary leg that had been grafted at 2½ days of age. *B* A 12-day-old chick embryo with a right supernumerary leg. The normal left leg is not visible. *C* Cross-section through the lumbar segment of the spinal cord of a chick with a supernumerary leg on the right-hand side. Note the increased number of neurons in the motor column (MC) and the dorsal root ganglion (DRG). *D* Counts of the number of neurons in the motor column in spinal cord sections running from rostral to caudal in chicks with supernumerary limbs. Note the higher cell counts on the experimental side. [From Hollyday M Hamburger V (1976) Reduction of the naturally occurring motor neuron loss by enlargement of the periphery. *J Comp Neurol* 170: 311–320.]

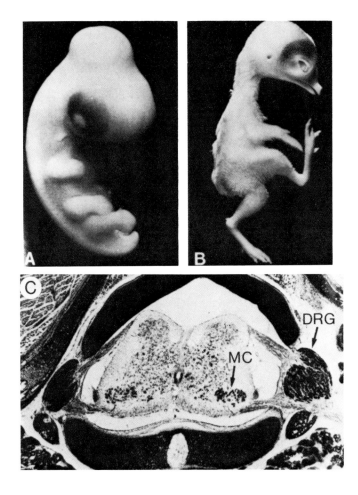

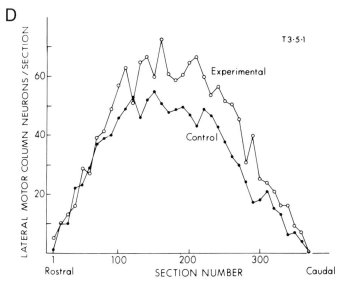

of this work was carried out in the developing chick. Through careful quantitative studies, Hamburger and his colleagues were able to show that in a number of regions, there was initially an overproliferation of cells followed by a period of naturally occurring cell death. Furthermore, manipulations of target tissue affected the extent of cell death rather than the initial proliferation.

For example, in the case of sensory ganglia, there are initially more cells in individual ganglia than will be found at maturity. Cell numbers are adjusted consequent to the death of some proportion of cells at a particular stage of development. Furthermore, the atrophy that occurs after removal of target tissue results from an increase in the extent of cell death during the period that it would normally occur. Similar mechanisms operate in the development of the spinal cord. The width of the lateral motor column initially is constant throughout the length of the spinal cord. As development proceeds, however, the column thins out in segments of the cord that are not associated with the limbs, that is, in all parts of the cord except the brachial and lumbar segments (Fig. 5.14). Hamburger found that during this period, the total number of neurons in the lateral motor column decreased by about 40% (Fig. 5.14).

An important feature of naturally occuring cell death is that it occurs during a defined temporal window; for some systems the period of cell death commences at approximately the same time that axons reach their targets. Thus, for a period of time, neurons develop independently of their targets; however, they become dependent upon their targets for survival at about the time that contacts are first made.

Cell death also seems to involve competition for available targets. The experiments that involve deleting or increasing the available postsynaptic targets can be interpreted in this way. Other experiments that support this interpretation involve manipulating the number of neurons in the presynaptic pool that are competing for the available postsynaptic targets. Thus, for example, the death of some populations of ciliary ganglion cells is reduced when other ganglion cells are prevented from contacting their

▷

FIGURE 5.14. Preferential sparing of motor neurons in brachial and lumbar segments of the spinal cord during naturally occurring cell death in the chick embryo. During normal development, the number of neurons in the motor column increases as cells migrate into the zone and then declines during the course of naturally occurring cell death (A–C). Cell death is most extensive in portions of the spinal cord that do not innervate limbs and less in the brachial and lumbar segments of the spinal cord that give rise to the innervation of the wing and leg. D Time course of naturally occurring cell death. CE = cervical segment; BR = brachial segment; TH = thoracic segment; L = lumbar segment; S = sacral segment. PVS = paravertebral sympathetic chain; DRG = dorsal root ganglion; VM = visceromotor neurons. [From Hamburger V (1975) Cell death in the development of the lateral motor column of the chick embryo. *J Comp Neurol* 160: 535–546.]

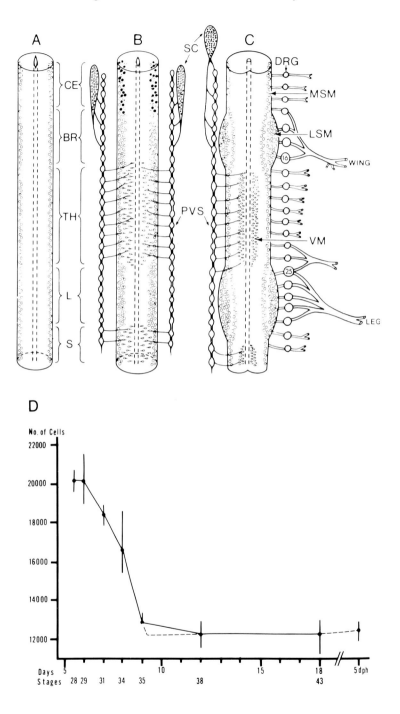

targets. Cell death in the isthmo-optic nuclei, which in the chick provide centripetal input to the eyes, is increased when the nuclei on both sides of the brain are forced to innervate a single eye following unilateral enucleation. Thus, reducing the amount of target tissue available for some number of presynaptic cells, or increasing the number of presynaptic cells competing for available targets, increases cell death in the presynaptic population. Alternatively, increasing the available target, or reducing the number of presynaptic cells competing for the available target, reduces cell death.

Taken together, these observations suggest that presynaptic neurons somehow compete with one another for the available target tissue, and cells that are not successful in this competition are eliminated. An attractive hypothesis is that cells compete for some factor provided by the postsynaptic target. This seems to be the case for at least some neurons. This conclusion is based on studies of nerve growth factor (NGF).

Cell Death, Trophic Factors, and the NGF Story

The line of research that led to the discovery and characterization of NFG was initiated by the finding that implantation of a tumor (a mouse sarcoma) into chick embryos induced a hypertrophy of nearby sensory ganglia. For a discussion of the history of NGF research, see Levi-Montalcini (1987). Following up on this work, Rita Levi-Montalcini and Hamburger (1951) found that these tumors induced a prodigious growth of both sympathetic and sensory fibers. Because the hypertrophy occurred when the tumor was placed on the chorioallantoic membrane rather than in the embryo, it was suggested that the effect was mediated by a diffusible substance. This agent was termed nerve growth factor. Thus began a line of investigation that would eventually lead to the Nobel Prize for Levi-Montalcini in 1986, which she shared with her former colleague Stanley Cohen who had worked on NGF early and then moved on to discover and characterize epidermal growth factor (EGF).

There have been a number of important steps in solving the NGF puzzle. A key step was the development of a bioassay for the factor. This assay took advantage of the effects that NGF exerted on dorsal root ganglia in culture (Fig. 5.15). Another crucial discovery was that some types of neurons depended on NGF for their survival. Treatment of mice with antibodies to NGF led to the disappearance of sympathetic neurons (Fig. 5.15). Moreover, NGF markedly promotes the survival of sympathetic neurons in culture. These studies formed the basis for the hypothesis that NGF regulates the extent of neuron survival in the sympathetic nervous system. The purification and sequencing of the NGF molecule was made possible by the fact that some tissues, particularly the submaxillary glands of mice, contain large amounts of NGF.

Other evidence suggested that NGF functioned as a trophic factor that is derived from target tissue. First, it was found that NGF is taken up by

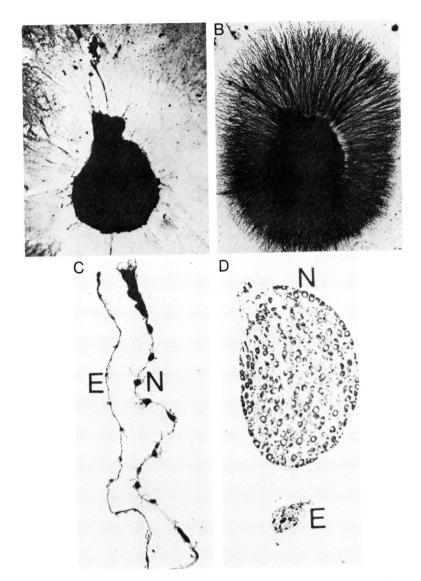

FIGURE 5.15. Effect of NGF on neurite outgrowth from sensory ganglia from a 7-day-old chick embryo in culture. *A* Control; *B* a comparable ganglion cultured in a medium supplemented with NGF (0.01 μg/mL) for 24 hours. *C, D* Effects of antisera to NGF. *C* Sympathetic chains from 1-month-old mice treated for five days after birth with an antiserum to NGF (E) compared with the sympathetic chain from a normal mouse (N). *D* Cross-section through the superior cervical ganglion of a mouse treated for three days after birth with an antiserum to NGF (E) compared with a ganglion from a normal mouse (N). Note the severe atrophy of ganglia from mice treated with the antiserum to NGF. [From Levi-Montalcini, R (1964) Growth control of nerve cells by a protein factor and its antiserum. *Science* 143: 105–110. Copyright 1964 by the AAAS.]

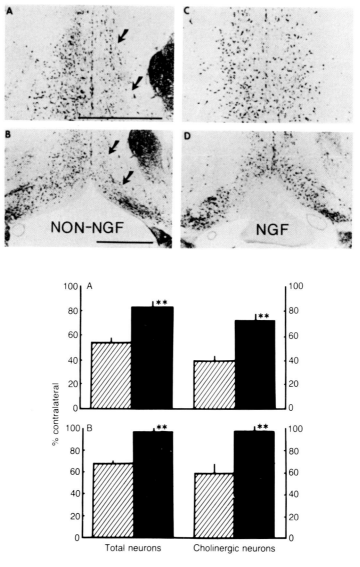

FIGURE 5.16. Reduction of retrograde degeneration in the CNS by exogenous NGF. Cholinergic neurons in the basal forebrain degenerate when their axons are cut. For example, transection of the hippocampal fimbria cuts the axons of the cholinergic neurons of the medial septal nucleus that project to the hippocampus. Many of these neurons degenerate following the damage, as seen on the right-hand side of *A* and *B* (*arrows*). The cell death is less extensive, however, when animals receive intraventricular infusions of NGF (*C* and *D*). Counts of the total number of neurons and the total number of cholinergic neurons reveal the extent of the sparing (lower portion of the figure). A larger percentage of cells survive in the NGF-treated animals (*solid bars*) than in nontreated animals (*hatched bars*). [From Williams LR Varon S Peterson GM et al. (1986) Continuous infusion of nerve growth factor prevents basal forebrain neuronal death after fimbria fornix transection. *Proc Natl Acad Sci USA* 83: 9231–9235.]

the terminals of neurons that were NGF-sensitive and transported retrogradely to the cell body. Second, NGF was found to be present in the
tissues that are targets for sympathetic axons; furthermore, the amount
of NGF correlated with the density of innervation. Third, exogenously
administered NGF was found to promote the survival of sympathetic neurons that were separated from their targets by axotomy. In addition, exogenous NGF could reduce the amount of naturally occurring cell death.
These latter observations suggested that NGF could substitute for the
survival factor normally provided by the target tissue. Recent studies have
revealed that some populations of CNS neurons also are sensitive to NGF
and that these cells can be spared from retrograde degeneration by the
administration of NGF (Fig. 5.16).

It is now accepted that NGF is produced by the targets of NGF-sensitive
neurons (sympathetic and sensory neurons in the periphery and some
populations of CNS neurons, including the cholinergic cells of the basal
forebrain). The NGF-sensitive cells have specific NGF receptors on their
axon terminals that take up NGF from the target cell. The NGF is then
transported back to the neuronal cell bodies, where it promotes cell survival and differentiation. In the absence of NGF, the neurons die. Thus,
in these systems, target-dependent cell death is regulated by the availability
of the trophic factor. As will be seen in Chapter 6, NGF may play other
roles in the regulation of axon growth.

Currently, it is not known if there are other trophic substances that
operate on other cell types in a manner similar to that of NGF. It is assumed, however, that such molecules exist, and several research groups
throughout the world are actively seeking such molecules. Certainly, the
discovery of other trophic substances is likely to have a major impact on
our understanding of neuronal interactions.

Cell Death and Functional Activity

A final point about target-regulated cell death during normal development
is that the death can be affected by functional activity. For example, the
death of cells in the lateral motor column of the chick is delayed when
neuromuscular transmission is chronically blocked (Fig. 5.17). When activity is restored, cell death recommences. These results are of interest,
since neuromuscular activity is also important for the elimination of multiple innervation during normal development (see Chapter 7).

Afferent Regulation of Cell Death

The wealth of evidence discussed above leaves no doubt that neuron
number is regulated, in part, as a result of interactions between neurons
and their targets. There is also evidence that cell survival can be regulated
by afferent innervation. The best example of such afferent regulation of
cell survival is found in the chick auditory system. Remarkably, the initial

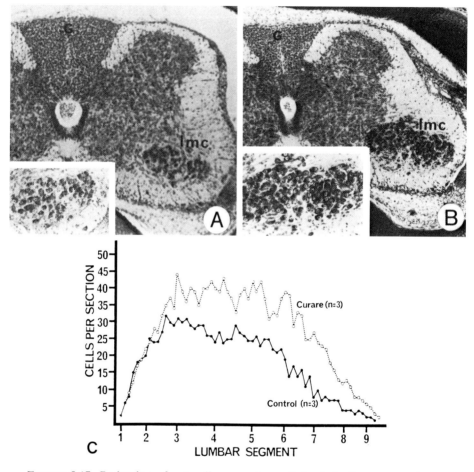

FIGURE 5.17. Reduction of naturally occurring cell death in the motor column when neuromuscular transmission is blocked with curare. *A* Cross-section through the lumbar cord of a normal 10-day-old chick; *B* Similar section through the lumbar cord of a chick that had been treated with curare between embryonic days 6 and 9. *C* Counts of the number of neurons in the motor column in spinal cord sections from lumbar segments 1–9. Note the greater number of neurons in the motor column in the chick treated with curare. [From Pittman R, Oppenheim RW (1979) Cell death of motoneurons in the chick embryo spinal cord. IV. Evidence that a functional neuromuscular interaction is involved in the regulation of naturally occurring cell death and the stabilization of synapses. *J Comp Neurol* 187: 425–446.]

studies in this system were carried out by Rita Levi-Montalcini in a primitive laboratory in her home in Italy during World War II when she and her mentor G. Levi were barred from holding academic appointments. Levi-Montalcini found that the elimination of innervation from the ear led to an increase in cell death in the brainstem nuclei receiving direct eighth nerve projections (Levi-Montalcini 1949). Subsequent studies have confirmed these results and have revealed that cell death also occurs when the eighth nerve is interrupted in posthatch chicks (Fig. 5.18). As is also true for target-mediated cell death, cells develop independently of their afferents until the initial contacts are formed. Interestingly, the cells become less sensitive to the removal of afferents during the late phase of development.

The above evidence makes it clear that during certain phases of development, the removal of afferent innervation can increase cell death. It is not known, however, whether naturally occurring cell death is actually regulated by afferent innervation. If it were possible to show that naturally occurring cell death occurs in parallel with a modulation of innervation, or that increased innervation prevents cell death, this hypothesis would gain strength. Meanwhile, it can only be said that during certain phases of development, some cells depend on afferent innervation for their survival.

In sum, the assembly of the nervous system involves highly regulated processes of cellular proliferation and migration that are followed by a reduction of cell numbers through programmed cell death. The latter process is regulated partly through the connections that neurons form with

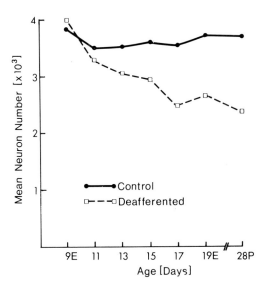

FIGURE 5.18. Afferent influences on cell death in the avian homologue of the anterior ventral cochlear nucleus after destruction of the cochlea during early development. The otocyst from which the cochlea develops was destroyed at 55 to 60 hours of incubation. The graphs indicate the average number of neurons in the nucleus on experimental and control sides at various ages. [After Parks TN (1979) Afferent influences on the development of the brain stem auditory nuclei of the chicken: otocyst ablation. *J Comp Neurol* 183: 665–678.]

one another. Thus, nervous system histogenesis really cannot be understood without a consideration of how the interconnections between neurons are formed. This topic will be considered in the next two chapters.

Supplemental Reading

General

Hamburger V (1981) Historical landmarks in neurogenesis. *Trends Neurosci* 4:151–155

Jacobson M (1978) *Developmental Neurobiology*. New York, Plenum Press

Purves D, Lichtman JW (1985) *Principles of Neural Development*. Sunderland MA, Sinauer Assoc Inc

Spemann H (1938) *Embryonic Development and Induction*. New Haven, Yale University Press

Spitzer NC (ed) (1982) *Neuronal Development*. New York, Plenum Press

Neurogenesis

Bayer SA, Yackel JW, Puri PS (1982) Neurons in the rat dentate gyrus granular layer substantially increase during juvenile and adult life. *Science* 216:890–892

Hirose G, Jacobson M (1979) Clonal organization of the nervous system of the frog. I. Clones stemming from individual blastomeres of the 16-cell and earlier stages. *Dev Biol* 71:191–202

Jacobson M, Hirose G (1981) Clonal organization of the central nervous system of the frog. II. Clones stemming from individual blastomeres of the 32- and 64-cell stages. *J Neurosci* 1:271–284

Determination of Differentiation

Furshpan EJ, MacLeish PR, O'Lague PH et al. (1976) Chemical transmission between rat sympathetic neurons and cardiac myocytes developing in microcultures: evidence for cholinergic, adrenergic, and dual-function neurons. *Proc Natl Acad Sci USA* 73:4225–4229

LeDouarin NM, Teillet M-A (1974) Experimental analysis of the migration and differentiation of neuroblasts of the autonomic nervous system and of the neuroectodermal mesenchymal derivatives using a biological cell marking technique. *Dev Biol* 41:162–184

Levitt P, Cooper ML, Rakic P (1981) Coexistence of neuronal and glial precursor cells in the cerebral ventricular zone of the fetal monkey: an ultrastructural immunoperoxidase study. *J Neurosci* 1:27–39

Patterson PH (1978) Environmental determination of autonomic neurotransmitter functions. *Ann Rev Neurosci* 1:1–17

Patterson PH, Chun LLY (1977) The induction of acetylcholine synthesis in primary cultures of dissociated rat sympathetic neurons. I. Effects of conditioned medium. *Dev Biol* 56:263–280

Migration

Rakic P (1971) Neuron-glia relationship during granule cell migration in the developing cerebellar cortex. A Golgi and electronmicroscopic study in *Macacus rhesus*. *J Comp Neurol* 141:283–312

Rakic P (1972) Mode of cell migration to superficial layers of fetal monkey neocortex. *J Comp Neurol* 145:61–84

Rakic P (1974) Neurons in rhesus monkey visual cortex: Systematic relation between time of origin and eventual disposition. *Science* 183:425–427

Rakic P (1981) Neuronal-glial interaction during brain development. *Trends Neurosci* 4:184–187

Target Regulation of Cell Death

Detwiler SR (1920) On the hyperplasia of nerve centers resulting from excessive peripheral loading. *Proc Natl Acad Sci USA* 6:96–101

Detwiler SR (1936) *Neuroembryology: An Experimental Study*. New York Macmillan

Hamburger V (1939) Motor and sensory hyperplasia following limb-bud transplantations in chick embryos. *Physiol Zool* 80:347–389

Hamburger V (1975) Cell death in the development of the lateral motor column of the chick embryo. *J Comp Neurol* 160:535–546

Hamburger V (1980) Trophic interactions in neurogenesis: A personal historical account. *Ann Rev Neurosci* 3:269–278

Hollyday M, Hamburger V (1976) Reduction of the naturally occurring motor neuron loss by enlargement of the periphery. *J Comp Neurol* 170:311–320

Oppenheim RW, Majors-Willard C (1978) Neuronal cell death in the brachial spinal cord of the chick is unrelated to the loss of polyneuronal innervation in wing muscle. *Brain Res* 154:148–152

Oppenheim RW, Chu-Wang I-W (1983) Aspects of naturally-occurring motoneuron death in the chick spinal cord during embryonic development, in Burnstock G, Vrbova G (eds): *Somatic and Autonomic Nerve-Muscle Interactions*. Amsterdam, Elsevier, pp 57–107

Shorey ML (1909) The effect of the destruction of peripheral areas on the differentiation of the neuroblasts. *J Exp Zool* 7:25–63

Nerve Growth Factor

Angeletti RH, Bradshaw RA (1971) Nerve growth factor from mouse submaxillary gland: amino acid sequence. *Proc Natl Acad Sci USA* 68:2417–2420

Berg DK (1982) Cell death in neuronal development. Regulation by trophic factors, in Spitzer NC (ed): *Neuronal Development*. New York Plenum Press, pp 297–331

Davies AM, Bandtlow C, Heumann R, Korsching S, Rohrer H, Thoenen H (1987) Timing and site of nerve growth factor synthesis in developing skin in relation to innervation and expression of the receptor. *Nature* 326:353–358

Hamburger V, Brunso-Bechtold JK, Yip JW (1981) Neuronal death in the spinal ganglia of the chick embryo and its reduction by nerve growth factor. *J Neurosci* 1:60–71

Hamburger V, Yip JW (1984) Reduction of experimentally induced neuronal death in spinal ganglia of the chick embryo by nerve growth factor. *J Neurosci* 4:767–774

Korsching S (1986) The role of nerve growth factor in the CNS. *Trends Neurosci* 9:570–573

Levi-Montalcini R, Hamburger V (1951) Selective growth-stimulating effects of mouse sarcoma on the sensory and sympathetic nervous system of the chick embryo. *J Exp Zool* 116:321–361

Levi-Montalcini R (1987) The nerve growth factor 35 years later. *Science* 237:1154–1162

Levi-Montalcini R, and Booker B (1960) Destruction of the sympathetic ganglia in mammals by an antiserum to a nerve growth protein. *Proc Natl Acad Sci USA* 46:384–391

Levi-Montalcini R, Cohen S (1956) *In vitro,* and *in vivo* effects of a nerve growth-stimulating agent isolated from snake venom. *Proc Natl Acad Sci USA* 42:695–699

Purves D (1986) The trophic theory of neural connections. *Trends Neurosci* 9/10:486–488

Afferent Regulation of Cell Death

Levi-Montalcini R (1949) The development of the acoustico-vestibular centers in the chick embryo in the absence of afferent root fiber and of descending fiber tracts. *J Comp Neurol* 91:209–241

Okado N, Oppenheim RW (1984) Cell death of motoneurons in the chick embryo spinal cord. IX. The loss of motoneurons following removal of afferent inputs. *J Neurosci* 4:1639–1652

Parks TN (1979) Afferent influences on the development of the brainstem auditory nuclei of the chicken: Otocyst ablation. *J Comp Neurol* 183:665–678

Other Neuronal Growth Factors

Thoenen H, Korsching S, Barde Y-A et al. (1983b) Quantitation and purification of neurotrophic molecules. *Cold Spring Harbor Symp Quant Biol* 48:679–684

Barde Y-A, Edgar D, Thoenen H (1983) New neurotrophic factors. *Ann Rev Physiol* 45:601–612

Cell Adhesion Molecules

Edelman GM (1986) Cell adhesion molecules in the regulation of animal form and tissue pattern. *Ann Rev Cell Biol* 2:81–116

Edelman GM, Crossin KL The molecular regulation of neural morphogenesis, in Easter SS, Barald KF, Carlson BM (eds): *From Message to Mind.* Directions in Developmental Neurobiology. Sunderland MA, Sinauer Associates, pp 4–22

Martini R, Schachner M (1986) Immunoelectron microscopic localization of neural cell adhesion molecules (L1, N-CAM, and MAG) and their shared carbohydrate epitope and myelin basic protein in developing sciatic nerve. *J Cell Biol* 103:2439–48

CHAPTER 6

The Formation of Neural Connections I: Axon Growth and Guidance

Introduction

For its proper operation, the nervous system depends on extremely specific interconnections between neurons, and between neurons and peripheral end-organs. Obviously, the specificity of connections requires a high degree of control during development. The initial steps in the establishment of connections involve the formation and proper positioning of the neurons (histogenesis, see Chapter 5). Next, the neurons must elaborate their axons and dendrites and form connections between the two. This chapter, together with Chapter 7, considers the mechanisms of the latter processes.

Elaboration of Axons and Dendrites

The formation of specific connections requires that the axonal process of the presynaptic cell comes into contact with the appropriate postsynaptic element. The importance of this process is well captured by Marcus Jacobson in his book on developmental neurobiology (Jacobson 1978): "The single major feature that distinguishes nerve cell differentiation and growth from that of all other types of cells is outgrowth of the axon from the nerve cell body in a specific direction, along a specific pathway, to form synaptic connections with specific targets. The invariance of the direction of initial outgrowth of the axon, its trajectory, and its targeting on postsynaptic sites are the *sine qua non* of axonal development." One might add that the formation of interconnections between CNS neurons also depends on the elaboration of the postsynaptic cell's receptive surface (dendrites).

In principle, the growth of axons to specific locations and the elaboration of a dendritic tree of a particular size and orientation could occur in one of two ways: (1) axons and dendrites could emerge from the cell randomly, with subsequent growth being regulated through interactions between the growing neurites and the tissue environment, or (2) the specificity of growth could be regulated by factors intrinsic to the neurons, such that the neuron would be programmed to grow axons and dendrites of a particular length and branching pattern in a particular direction with respect to the cell body. Not surprisingly, given the very different structural properties of axons and dendrites, the growth of axons versus dendrites is regulated in different ways. In general *the growth of axons is determined largely by interactions between the growing axon and the tissue environment, whereas the elaboration of dendrites depends mostly on factors that are intrinsic to the cell of origin.* (See Chapter 5 for a discussion of the terms "intrinsic" versus "extrinsic" as they pertain to developmental issues.) These statements, like most generalizations, are not absolute; nevertheless, they provide a useful starting point. This chapter and Chapter 7 will consider these generalizations and their exceptions in some detail, with this chapter focusing on the regulation of axonal growth and Chapter 7 considering the growth of dendrites and the formation of synaptic connections.

The generalizations above emphasize the differences between axons and dendrites; however, at least two aspects of growth are similar in axons and dendrites. These are (1) the mechanisms controlling the emergence of neurites from an undifferentiated cell body and (2) the mechanisms of membrane addition in the growing process.

The Emergence of Axons and Dendrites from the Cell Body

It is thought that factors intrinsic to the cell of origin determine where *neurites* (the general term that applies to both axons and dendrites) emerge from the cell body. In addition, the number of processes is thought to be intrinsically regulated. This conclusion is based on several observations. First, neurons typically give rise to only one axon and a predictable number of dendrites based on the cell type. In addition, the axon and dendrites appear to originate from predictable locations on the cell body. This fact is well illustrated by pyramidal cells of the cerebral cortex. The main dendritic trunk of cortical pyramidal cells originates from the apex of the pyramid-shaped cell body. The axon and basal dendrites emerge from the basal pole. Most pyramidal neurons are aligned within the cortex so that their apical dendrite points toward the cortical surface and the axon toward the white matter. However, sometimes cortical neurons are misaligned, presumably because they do not complete a full 180° rotation after migration. Even when misaligned, cortical neurons still give rise to processes that are identifiable as apical and basal dendrites, together with a single axon that usually emerges from the base of the pyramid (Fig. 6.1). Thus, pyramidal cells that are completely inverted give rise to an axon that emerges from the base of the pyramid (and thus begins to grow toward the cortical surface), whereas the apical dendrite grows in the opposite direction, toward the white matter. These observations have been taken

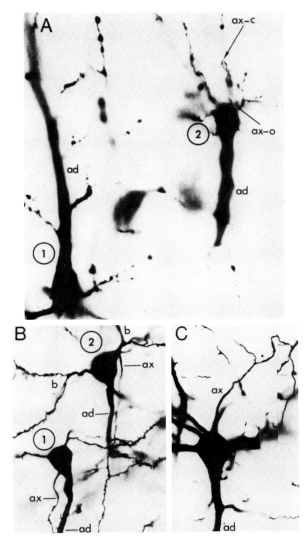

FIGURE 6.1. Inverted pyramidal cells in the cerebral cortex of the rabbit. *A* Normally oriented (1) and inverted (2) pyramidal cells. The axon of cell 2 emerges from the base of the pyramidal cell and initially grows toward the pial surface, making a hairpin turn after growing for a few microns (ax-c). The apical dendrite (ad) is oriented toward the white matter. *B* Two inverted pyramidal cells in the occipital cortex of an adult rabbit. In these cells the axon originates from abnormal locations. The axon of cell 1 originates from a point on the cell body near the apical dendrite, whereas the axon of cell 2 originates from a point near one of the basal dendrites. In both of these cells the orientation of the apical dendrite is appropriate with respect to the cell body. *C* An inverted pyramidal cell with an axon that emerges from the base of the pyramidal cell body. The axon grows initially toward the pial surface, but then reorients to grow toward the white matter. [From Van der Loos H (1965) The "improperly" oriented pyramidal cell in the cerebral cortex and its possible bearing on problems of growth and cell orientation. *Bull Johns Hopkins Hosp* 117: 228–250.]

as evidence that the point of origin of axons and dendrites, as well as the direction of their initial outgrowth, are determined by factors that are intrinsic to the neuronal cell body.

How the point of origin of the different types of neuronal processes is regulated is not known. On the basis of studies of other cell types, it is thought that the neuronal cytoskeleton may have some form of organizing center that leads to the elaboration of processes at fixed locations with respect to the internal architecture of the neuron. In fact, it is thought that the polarization of the cell is fixed at the time of the last mitosis. Evidence in support of this contention comes from studies of neuroblastoma cells grown in culture, where sister cells arising from a final division often develop processes that are mirror images of one another. It is also of interest that stacks of Golgi apparatus are often found at the base of both axons and dendrites, and these organelles could represent part of the cellular apparatus that produces a dendrite or axon at a particular location. In fact, Ramon y Cajal observed that the Golgi apparatus is very prominent beneath the portion of the cell from which both axons and dendrites arise. While these observations are provocative, it is not clear how the polarized distribution of organelles is actually regulated.

Although the point of origin of both axons and dendrites seems to be determined by intrinsic factors, the two types of processes exhibit different patterns of outgrowth after leaving the vicinity of the cell body. Before considering how growth and guidance are regulated, it is important to consider the general issue of how neuronal processes elongate.

The Elongation of Neuronal Processes

The elongation of axons and dendrites requires that new membrane be added to the growing process together with additional cytoplasmic and cytoskeletal material. There are essentially three ways that such growth could occur: (1) Neurites could grow at the base like hair, in which case new membrane and additional cytoplasmic constituents would simply be added at the base. (2) New membrane could be added throughout the length of the neurite (ie, growth could be intrasegmental), in which case cytoplasmic and cytoskeletal constitutents would still have to be added at the base. (3) New membrane could be added at the tip. In this case, elongation would occur as a result of the construction of a new segment; again, however, cytoplasmic and cytoskeletal constituents would have to be added at the base. The distinction betweeen these modes of growth is important, since the mechanism places constraints on the types of regulatory processes that could be exerted during growth.

With axons the evidence is rather convincing that elongation occurs as a result of the addition of membrane at the tip. The most direct evidence comes from studies of the branching pattern of axons growing in culture. If axons grew by adding membrane along their length (i.e., *intrasegmen-*

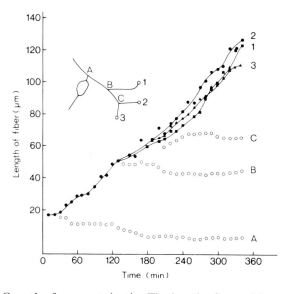

FIGURE 6.2. Growth of axons at the tip. The length of axonal branches of single neurons was measured by time-lapse cinematography in tissue culture. There was a continuous increase in axon length over time. However, once a branch was formed, the distance from the cell body to the branch did not increase. This indicates that elongation occurs as a result of the addition of material at the tip of the growing axon. [Reproduced from *The Journal of Cell Biology* 1973; 56: 702–712, by copyright permission of the Rockefeller University Press.]

tally) or by adding membrane at the base, then the segment length between branches should increase as the axon elongates. However, studies of neurons growing in culture reveal that the distance between branches remains constant, and elongation of the axon occurs exclusively as a result of the elongation of the tip of the axon (Fig. 6.2). This does not imply that axons and dendrites cannot elongate intrasegmentally. In fact, axons can be "stretched" or "towed" after establishing their connections. It is in this manner that axons of the peripheral nervous system can accommodate for body growth during development. Similarly, intrasegmental growth of dendrites probably occurs during the development of cortical structures to accommodate for the increase in the thickness of the cortex as it matures. This type of elongation may simply represent an accommodation to some stretching force rather than an active growth process. In any case, it is clear that the predominant form of growth involves the addition of new membrane at the tip of the growing neurite.

The Axonal Growth Cone

The growing tips of axons and dendrites are marked by highly motile structural specializations termed *growth cones*. Axonal growth cones are generally flat, hand-shaped structures that give rise to several thin filo-podial extensions (Fig. 6.3). The motility of growth cones has been amply documented by studying neurons in tissue culture and by studying growing axons in the transparent tails of tadpoles. These studies reveal that the

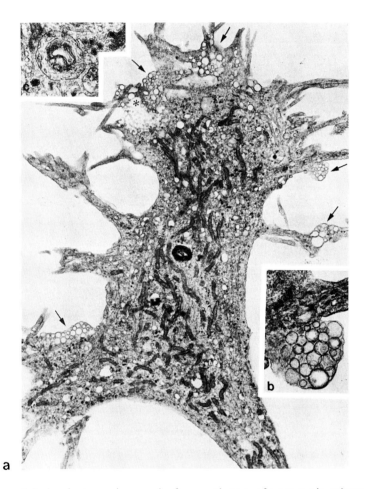

FIGURE 6.3. An electron micrograph of a growth cone of a neuron in culture. Note the numerous filopodia and the vesicle-filled mounds (*arrows*). Filaments and mi-crotubules are present throughout the growth cone, although they are difficult to see at this magnification. The granular cytoskeleton in the edges of the growth cone consists of an actin matrix. Mitochondria and a large multivesicular body (MVB) are also present. *b* An enlargement illustrating one of the vesicle-filled mounds. *c* An enlargement illustrating a multivesicular body with a membranous whorl termed a "myelin figure." The magnification in $a = 11,000$; $b = 39,000$; $c = 34,000$. [From Bunge MB (1977) Initial endocytosis of peroxidase or ferritin by growth cones of cultured nerve cells. *J Neurocytol* 6: 407–439. Reprinted by permission of Chapman & Hall, Publishers.]

filopodia of growth cones extend and retract at a rate of about 6 to 10 μm/min. The membrane around the flattened growth cone exhibits "ruffling" very much like that of the leading edge of a migrating fibroblast. The filopodia and the area underlying the ruffling membrane are filled with microfilaments. Immunocytochemical studies reveal that these microfilaments are composed of actin.

In addition to actin filaments, other specializations of the growth cone include vesicles. These vesicles may represent newly synthesized membrane that will fuse with the plasma membrane of the growth cone or membrane that has been retrieved from withdrawing filopodia. Many of the vesicle-filled mounds are thought to be artifactual, coming about during fixation. Microtubules and neurofilaments of the axon extend partially into the growth cone and sometimes splay out into the flattened central portion. However, microtubules and neurofilaments do not extend into the filopodia or into areas underlying the ruffling membranes.

Intrinsic Versus Extrinsic Controls on Axon Growth

Shortly after leaving the immediate vicinity of the cell body of origin, growing axons seem to be guided entirely by interactions with their cellular environment. This guidance is thought to come about through interactions of the growth cone with elements that it encounters. For example, in the case of the misoriented pyramidal cells mentioned above, axons that begin to grow toward the cortical surface quickly reverse direction and grow toward the white matter (Fig. 6.1). Another example that is often cited involves Mauthner cells. In salamanders, pieces of brainstem including Mauthner cells can be transplanted; when the transplants are reversed, the axons initially grow in a rostral direction, but reverse direction near the graft-host boundary and then grow caudally (Fig. 6.4).

The mechanisms of axon guidance seem to involve interactions between axonal growth cones and the tissue environment. Studies of axons in tissue culture reveal that the growth cones seem to explore their immediate cellular environment with their filopodia. When the filopodia encounter unfavorable terrain, they withdraw. When they encounter favorable terrain, the filopodia remain, and the axon grows along that terrain.

In the organism, axons grow for long distances, bypassing other neurons, intersecting other fiber pathways, and taking entirely predictable routes through their cellular terrain. Understanding the development of specific connections depends on an understanding of how these axons are guided.

Axonal Guidance

A variety of extrinsic factors influence the course taken by growing axons. The best evidence involves direct manipulations of growing axons either in vivo or in vitro. However, whether all of these influences actually op-

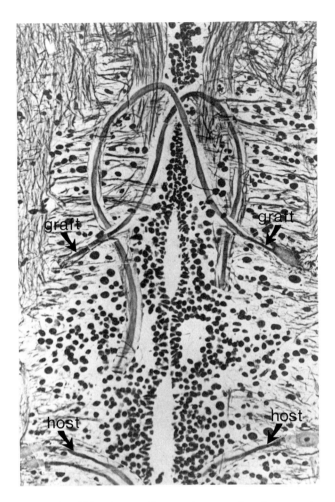

FIGURE 6.4. Growth of Mauthner cell axons in normal and reversed segments of the medulla. A segment of medulla was transplanted with a rostrocaudal inversion into the medulla of a host salamander. Mauthner neurons were present in the medulla of the host and in the graft. The axons of normal Mauthner cells of the host grew caudally; the axons of Mauthner cells of the graft grew rostrally for a short distance and then reversed direction to grow caudally. The rostral boundary of the graft is near the point of reversal, while the caudal boundary is just rostral to the Mauthner cells of the host. [From Hibbard E (1965) Orientation and directed growth of Mauthner cell axons from duplicated vestibular nerve roots. *Exp Neurol* 13: 289–301.]

erate in the living organism is not known; showing that a particular factor *can* have an effect does not necessarily indicate that such a factor does have that effect normally.

Stereotropism

One factor that can influence the direction of growth is mechanical. Axons tend to grow along surfaces. This tendency is termed *stereotropism, tactile adhesion,* or *contact guidance.* Early studies in tissue culture provided evidence for this effect, since axons tend to grow along scratches in culture dishes or along lines of stress induced by stretching plasma clots (the substance in which early tissue culture was carried out). A modern version of the hypothesis that axons are guided by mechanical cues is the *blueprint hypothesis;* this hypothesis holds that axons grow along the surfaces of ependymoglia through the spaces created by the neuroepithelial matrix (see Chapter 2 for a discussion of the possible role of radial glia in guiding the growth of long axon tracts). Of course, such guidance can only operate where there is such a matrix.

Although mechanical factors could provide some framework for axon growth, the highly specific growth exhibited by most axons cannot be explained entirely on this basis. For example, mechanical factors cannot explain how axons sort themselves out when fibers of different systems cross one another or intermingle, which is clearly quite common in the CNS. In addition, mechanical factors cannot explain preferred directions of elongation (either direction along a pathway should be equally favorable).

Tropism Based on Differential Adhesiveness

There is clear evidence that axons adhere better to some substrates than others and that they grow better along adhesive substrates than along nonadhesive ones. When a culture dish is coated in a gridlike fashion with an adhesive substrate (such as polyornithine, laminin, etc.), axons will grow preferentially along the adhesive substance, avoiding entirely areas where the adhesiveness is low (Fig. 6.5). There is little direct evidence for substrate preferences based on differential adhesiveness in vivo. However, axons do seem to follow preferred substrates; for example, some axons seem to follow basal lamina.

As was the case with mechanical factors, it is difficult to account for all aspects of the highly specific growth of CNS pathways by general preferences for particular substrates. For example, different fiber systems often intermix, but then separate to follow different routes. Thus, if the adhesive properties of different substrates are to account for specific growth, *different neurons must have different adhesive properties.* Furthermore, there must be some directionality to the adhesiveness in order to explain the directionally specific growth of CNS pathways. One possibility is that

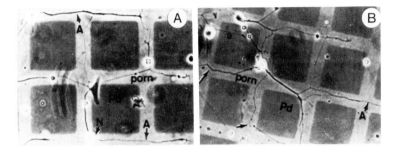

FIGURE 6.5. Preferential growth of axons along an adhesive substrate. Adhesive properties of various substrates were assayed by growing neurons on the substrates and evaluating the percentage of growth cones that were displaced by blasts of air. In this assay, collagen and polyornithine were found to be very adhesive substrates. When neurons were cultured in a dish where the substrate was coated with polyornithine in a gridlike configuration, growing axons extended only along the polyornithine (porn)-coated substrate. Similar results were obtained using collagen-coated substrates. N = nerve cell body; A = axon. [From Letourneau PC (1975) Cell-to substratum adhesion and guidance of axonal elongation. *Dev Biol* 44, 92–101.]

there could be gradients of adhesive substances so that axons grew toward areas of higher adhesivity.

Thus far there is not a great deal of direct evidence for different neurons adhering differentially to substrates, or for gradients in adhesiveness. Nevertheless, naturally occurring interactions between axons and between axons and substrates provide hints that such adhesiveness may occur. For example, a well-recognized phenomenon is for axons of a given type to fasciculate. The fasciculation of axons helps account for the existence of tracts in the CNS. Fasciculation might arise from the same kind of selection adhesion between axons that accounts for their growth along certain substrates. Indeed, studies of neurons grown in culture indicate that there is a reciprocal relationship between adhesion to a substrate and adhesion to other axons; reductions in the adhesivity of the substrate increase fasciculation.

Selective adhesion between classes of axons could be a result of complementary recognition molecules on the cell membranes such as the CAMs discussed in Chapter 5. The existence of complementary recognition molecules would also explain the reciprocal relationship between substrate adhesivity and fasciculation. When the adhesivity of the substrate is too high, these recognition molecules are essentially out-competed by the molecules of the substrate. In this way, the "selective adhesion hypothesis" is closely related to the "chemoaffinity hypothesis" for the formation of specific connections (see Chapter 7).

Galvanotropism

Early theories of axon guidance suggested that growing axons could be influenced by electrical fields. Although this hypothesis fell out of favor for a number of years, interest in it has been rekindled by recent studies that reveal voltage gradients in developing embryos and regenerating limbs and a sensitivity of growing axons to such fields. These fields are thought to be generated by local electrogenic membrane pumps, and fields of up to 100 mV have been measured. Application of electrical gradients of a few millivolts to axons growing in culture does influence the course of the axon; the axons grow toward the cathode. In general, although effects can be demonstrated, they are not dramatic, and they require fairly large fields. The probability seems remote that such fields could play a significant role in axonal navigation in vivo.

Chemotropism

Another popular hypothesis is that axons navigate on the basis of chemical cues (chemotropism). This hypothesis holds that neurons navigate on the basis of gradients of diffusible substances that are released from the target. By growing toward areas of higher concentrations of a *tropic substance,* axons could eventually reach the source (their targets). Note here the distinction between *tropic substances,* which attract growing axons, and *trophic substances,* which are required for the survival of cells or which stimulate cellular metabolism.

At the outset it is important to set forth the requirements for a chemotropic mechanism based on a target-derived factor. First, a diffusible chemotropic factor must be produced in the target and released into the extracellular medium. Second, it must be capable of affecting the direction of axon growth (specifically, axons should respond to gradients of the factor by growing toward the source). Third, the substance must be present at the appropriate time (during the period of axon growth); indeed, one would expect that the production of the factor should be highest during the time that the axon is growing toward the target and before the axon has actually made contact.

The best evidence for chemotropic effects on growing axons involves *nerve growth factor (NGF).* First, NGF is a soluble protein that is produced in the targets of sympathetic axons; it greatly enhances the growth of sympathetic axons and axons from the dorsal root ganglia (see Chapter 5). Second, gradients of NGF can influence the *course* taken by growing axons. For example, when NGF-sensitive neurons are grown in culture, their axons will actually grow toward an NGF source (eg, a micropipette containing NGF; see Fig. 6.6).

Interestingly, the effect of NGF on growing axons may involve a different mechanism than the *trophic* effect on the cell body. There is good

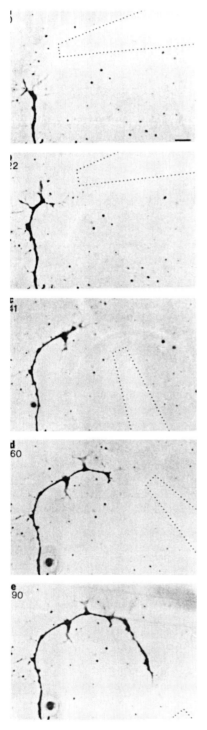

FIGURE 6.6. Guidance of growth cones by gradients of NGF. The figure shows sequential photographs of the growth cone of a dorsal root ganglion cell in culture. A pipette containing NGF was positioned to the side of the growth cone (*dotted outline*). The axon reorients and grows toward the pipette. Numbers indicate the time in minutes after the onset of perfusion via the micropipette. [From Gunderson RW, Barrett JN (1979) Neuronal chemotaxis: chick dorsal-root axons turn toward high concentrations of nerve growth factor. *Science* 206: 1079–1081. Copyright 1979 by the AAAS.]

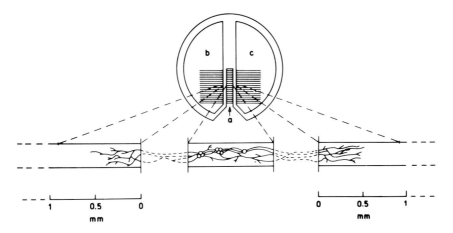

FIGURE 6.7. Axon growth is regulated by local NGF concentrations. Sympathetic neurons were plated in the center compartment of the chamber separated into three compartments by a Teflon divider. The collagen-coated floor of the dish was scored with parallel scratches that spanned the compartments, forming collagen channels along which the axons could grow. The neurites penetrated the silicone grease barriers beneath the Teflon dividers and extended into the side chambers. The NGF-dependent sympathetic neurons survived when NGF was present either in the center compartment or in one of the side compartments. Thus, provision of NGF to axons is sufficient to provide trophic support to an NGF-dependent cell body. NGF also exerts local effects, however, since the growth of neurites in side chambers with NGF was substantially greater than in side chambers without NGF. Thus, the extent of axon arborization is regulated by local NGF concentrations in the environment of the axon. [From Campenot RB (1982) Development of sympathetic neurons in compartmentalized cultures. I. Local control of neurite growth by nerve growth factor. *Dev Biol* 93: 1–12.]

evidence that trophic effect of NGF involves the uptake and retrograde transport of NGF to the cell body, where it affects gene expression. However, the growth of axons of NGF-sensitive neurons can be affected by local concentrations of NGF around the axons themselves. This has been demonstrated by growing NGF-sensitive neurons in chambers where the medium surrounding the axon is not in communication with the medium surrounding the cell body (Fig. 6.7). Such chambers allow independent manipulations of NGF concentrations around axons and around cell bodies. When sympathetic neurons are grown in the center compartment of such chambers, they send axons into the two side chambers, providing that NGF is present there. Furthermore, these NGF-sensitive neurons survive even when NGF is present only in the side chambers containing the axons. When NGF is present in only one of the side chambers, the axons arborize extensively within that chamber, but not in the opposite chamber without NGF. These results reveal that axon growth can be regulated by local concentrations of a trophic factor.

Although NGF is produced in the targets of NGF-sensitive neurons and can affect the course of axon growth, its potential role as a *diffusible* chemotropic factor has been seriously challenged by recent experiments by Davies et al. (1987), who evaluated the time course of NGF expression during development. In contrast to the expectations of the hypothesis that factors released by the target guide the axon, NGF is *not* produced to any great extent by the targets of sensory axons until *after* they have been contacted by the axons. Moreover, NGF receptors could not be detected on growing sensory neurons until *after* the growing axons reach their target. This timing suggests that the NGF that is produced by the target may play only a minimal role in guiding the growing axons to their target and more likely serves as a trophic factor to maintain the neurons after they form connections.

In general, even if chemotropic substances are produced and released by target cells, it seems likely that they could influence the course of axons only over relatively short distances. It certainly seems unlikely that gradients of diffusible substances could account for the long-distance navigation of axons over distances of millimeters or meters.

Chemical Markers for Routes of Axon Growth

Much of the initial growth of the axon is highly directed and specific, occurring along predictable pathways; this is true even when the axons are relatively distant from their targets. For example, motor and sensory neurons grow into limbs along highly consistent pathways. It is very unlikely that this patterned growth could be regulated by factors released from a distant target. A hypothesis related to the chemotropic hypothesis is that axons navigate by following chemical cues that are present along their route of growth (i.e., that there are routes with specific chemical labels).

The hypothesis that axons follow labeled routes makes several predictions: (1) that axons can follow routes with particular chemical labels, (2) that different neurons seek out different labels, and (3) that there are biochemical differences that mark different pathways in situ. There is some evidence to support these predictions. For example, using a strategy similar to that described above for studies of neuronal adhesiveness, it has been shown that axons do show preferences for certain substrates. For example, R.W. Gunderson (1985) has studied the growth of neurites of sensory neurons on substrates labeled with NGF. When NGF was bound to an otherwise uniform substrate layer in a grid pattern, the neurites of sensory neurons obtained from seven-day-old embryos followed the NGF-rich substrates very precisely and avoided the NGF-free regions (Fig. 6.8).

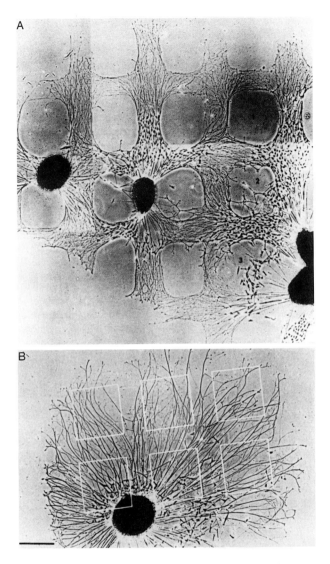

FIGURE 6.8. Selective growth of axons along NGF-coated substrates. NGF was bound to an otherwise uniform substrate in a gridlike configuration. *A* The outgrowth of axons from dorsal root ganglia from 7-day-old chick embryos occurred selectively along the portions of the substrate that had adsorbed NGF. The figure illustrates the pattern of neurite outgrowth after 72 hours in vitro. The areas that are free of adsorbed NGF are square in outline and are bounded by neurites. *B* Axons from dorsal root ganglia from 17 day old embryo do not show the selective growth; instead the axons grow readily across NGF-free areas. [From Gunderson RW (1985) Sensory neurite growth cone guidance by substrate adsorbed nerve growth factor. *J Neurosci Res* 13: 199–212.]

Interestingly, neurites from neurons harvested from 17-day-old embryos did not show the same patterned growth; instead, these neurites grew readily across NGF-free portions of the substrate. These observations reveal that axons can follow routes labeled with NGF and that neurons respond differentially to the labeled routes at different times of development.

There is currently a great deal of interest in the question of whether there are biochemical differences between pathways in vivo that might play a role in axon guidance. Although differences have been found, the challenge will be to show that the biochemical differences between pathways actually represent the mechanism of guidance. Furthermore, some aspects of neural growth do not seem to be in accord with the "labeled route" hypothesis. For example, axons can often be diverted from their normal routes, yet still eventually reach their targets.

Guideposts for Axonal Growth

Studies of axon growth in insects have revealed that axons not only follow precise routes, but also contact identifiable cells along their route of growth. These observations have led to the concept of "guideposts," where axons are thought to grow from one guidepost to the next along their course. In grasshoppers, for example, sensory axons growing into the limbs contact identifiable cells along their course, although the growing axon does not apparently form connections with these cells. When these "guidepost" cells are removed, the normal growth of the axon is disrupted. Whether similar guideposts exist in higher organisms is not clear. Again, fixed guideposts seem unlikely to account for the entire story, since axons can be diverted in their course and still come to reach their normal targets.

In the end it may well be that a number of factors interact to guide the axon from its point of origin to its target. Stereotropism could operate in a general way to keep growing fibers within bounds. Within these bounds there may be areas with differential adhesivity or labeled routes for particular axons to follow. Finally, chemotropism and/or galvanotropism may guide the axon to its postsynaptic site after it has reached the appropriate target area. It may well prove to be impossible to dissociate completely these various contributors.

The Problem of Distance and the Concept of Pioneer Fibers

It is also important to recall that the distance between interconnected structures in the mature animal does not necessarily indicate the actual distance that early-forming axons would have to grow. Early in development, distances are considerably shorter; in fact, some structures that eventually interconnect (the eye and brain for example) lie quite close to one another in the embryo. For some structures it may well be that a few "pioneer" connections are established quite early and that these axons become stretched as the development of the organism proceeds. Later-developing fibers could then simply follow these *pioneer fibers* as a result of selective fasciculation. Although there are as yet few well-documented examples of these kinds of phenomena, the hypothesis is hard to disprove, since very few connections need be present to act as pioneers.

The Establishment of Connectivity in Parallel with Migration

Although many pathways of the CNS must grow for some distance, some connections are established during the course of migration. The best example of this is found in the cerebellum. Granule cells are generated in the external granular layer and migrate through the molecular layer to their final position just deep to the Purkinje cell layer. As they pass through the molecular layer, they give rise to the parallel fibers. They then continue their migration, trailing their primary axon (see Chapter 5).

Axonal Overproliferation and Withdrawal

Specific growth is the rule in most systems, but some axonal projections are initially less specific than they will be in the mature organism. Indeed, some systems exhibit a considerable overproliferation of their axons, projecting to areas where they normally do not project in the mature animal. These "ectopic" projections are then withdrawn during the course of development.

There are several well characterized examples of overproliferation followed by pruning of ectopic projections. For example, many pathways that are predominantly unilateral in mature animals are bilateral early in development. Retinal projections are a case in point. In adult rats most retinal projections to the CNS are crossed; nevertheless, during early development there is a substantial ipsilateral projection that is eliminated as animals mature (Fig. 6.9). Interestingly, the elimination of these projections seems to depend on some competitive interaction between related projection systems. Thus, the atypical (for an adult) ipsilateral projections from the eye are retained if the opposite eye is removed. Furthermore, these competitive interactions seem to depend on activity: Blocking ganglion cell activity with the sodium channel blocker tetrodotoxin (TTX) has the same effect as removing the eye, in that the atypical ipsilateral projections from the opposite eye are retained.

Studies of the retinal projections of primates reveal that the elimination of projections may contribute to the development of normal specificity. In higher mammals, including primates, the projections from the retina

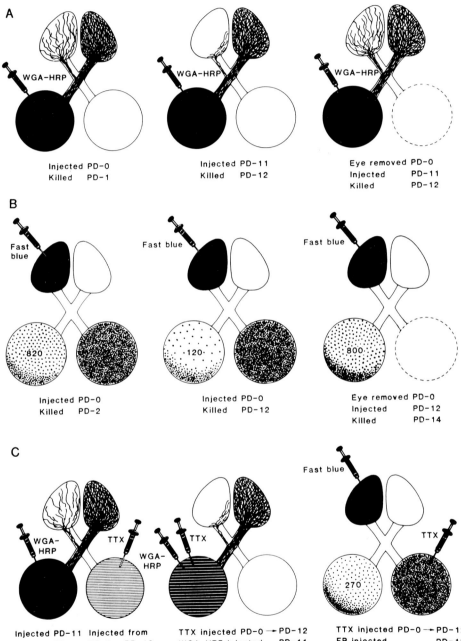

A

Injected PD-0
Killed   PD-1

Injected PD-11
Killed   PD-12

Eye removed PD-0
Injected   PD-11
Killed   PD-12

B

Fast blue

Injected PD-0
Killed   PD-2

Injected PD-0
Killed   PD-12

Eye removed PD-0
Injected   PD-12
Killed   PD-14

C

WGA-HRP   TTX   WGA-HRP

Injected PD-11 Injected from
Killed   PD-12 PD-0→PD-12

TTX

TTX injected PD-0 → PD-12
WGA-HRP injected   PD-11
Killed   PD-12

Fast blue   TTX

TTX injected PD-0 → PD-12
FB injected   PD-12
Killed   PD-14

to the dorsal lateral geniculate nucleus (LGd) are segregated into eye-specific laminae (Fig. 6.10). However, early in development the projections from the retina terminate throughout the LGd, with no evidence of segregation into laminae. Thus, the terminal arbors of axons from the two eyes are thought to segregate into nonoverlapping laminae after a period when overlap is extensive. Quantitative studies of the number of axons in the optic nerve of the monkey have revealed that there are actually more axons present early in development than are maintained in adulthood (Fig. 6.11). The elimination of axons in the optic nerve occurs during approximately the same period that the terminal fields of axons from the two eyes become segregated in the lateral geniculate nucleus (LGN). Thus, it is thought that the elimination of axons plays a role in establishing the normal pattern of innervation in the LGN, with the projection to the inappropriate laminae being eliminated as a consequence of axon withdrawal.

Other examples of overproduction and withdrawal of axons involve corticofugal pathways. In adult animals only a few cortical areas give rise to corticospinal projections. In young animals, however, widespread areas give rise to such projections; these are withdrawn as the animals mature. In addition, the pattern of interconnection between the two cerebral hemispheres is established by eliminating projections; as a result, a projection that is initially widespread becomes patchy.

The focusing of cortico-cortical projections involves a complete re-

◁──

FIGURE 6.9. Elimination of early widespread projections in the visual system. In rats the projections from the eye to the superior colliculus (SC) are predominantly contralateral, with the ipsilateral component of the pathway being restricted to the rostromedial portion of the SC. During early development, however, there is a substantial ipsilateral projection throughout the SC. This can be revealed by injecting orthograde tracers into the eye (in this case wheat germ agglutinin-horseradish peroxidase:WGA-HRP). When WGA-HRP is injected on the day of birth (PD-O), the ipsilateral component of the projection can be seen (A). The ipsilateral component disappears by PD-11. If one eye is removed on PD-O, however, the widespread projections of the remaining eye remain. B Retrograde labeling of the cells of origin of the retinal projections to the SC reveals that the restriction of the ipsilateral projection comes about as a result of the death of ganglion cells outside the inferior temporal retina. Injections of retrograde tracers (true blue) into the SC on PD-O labeled about 800 cells outside of the inferior temporal retina, whereas only 120 labeled cells were present in this zone by PD-12. Removal of the contralateral eye leads to the sparing of these neurons. This result is particularly important, since it demonstrates that the disappearance of labeled cells is not due to the loss of the label from cells that survive. C Blocking electrical activity in the contralateral retina prevents the restriction of the projections from the opposite eye and spares the cells of origin of the "ectopic" projections. [From Cowan WM, Fawcett JW, O'Leary DDM et al. (1984) Regressive events in neurogenesis. *Science* 225: 1258–1263. Copyright 1984 by the AAAS.]

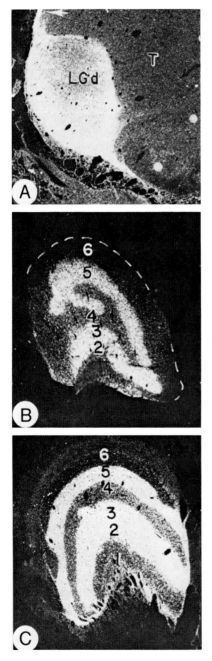

FIGURE 6.10. Development of segregated retinal input to the lateral geniculate nucleus. The projections from the eye to the dorsal lateral geniculate nucleus (LGd) were traced by injecting a mixture of 3H-proline and 3H-fucose into one eye of fetal monkeys. In mature monkeys the projections from each eye terminate in nonoverlapping laminae. This segregated pattern of termination develops progressively from a pattern where the terminal fields overlap. *A* Pattern of termination of the projections from one eye from an animal that had been injected on E64. The silver grains indicating labeled projections appear white in this dark-field photograph. Note the uniform pattern of labeling, with no evidence of segregated laminae. *B* Segregated pattern of termination is evident when eyes were injected at E110. *C* Pattern of termination at E144 is similar to that of mature monkeys. 3V = third ventricle. Arabic numerals denote layers in the LGd. [From Rakic P (1977) Prenatal development of the visual system in rhesus monkey. *Phil Trans R Soc Lond* (Biol) 278: 245–260.]

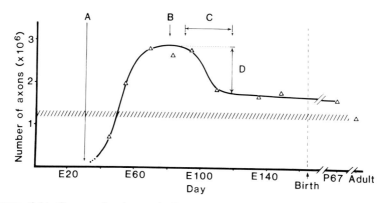

FIGURE 6.11. Overproduction and elimination of axons in the optic nerve during early development. Counts were made of the total number of axons in the optic nerve of rhesus monkeys of various embryonic and postnatal ages. *Striped line* indicates the number of optic axons in adult monkeys. *Arrow* in *A* indicates the onset of the genesis of retinal ganglion cells (based upon 3H-thymidine labeling studies). *Arrow* in *B* indicates the end of the period of ganglion cell histogenesis. *C* Period when retinal input from the two eyes becomes segregated in the lateral geniculate nucleus. [From Rakic P, Riley KP (1983) Overproduction and elimination of retinal axons in the fetal rhesus monkey. *Science* 219: 1441–1444. Copyright 1983 by the AAAS.]

traction of the projections by some cells. This has been demonstrated by retrograde labeling with horseradish peroxidase (HRP). In young animals HRP injections lead to retrograde labeling of cells throughout the cerebral cortex contralateral to the injection. In mature animals, however, equivalent injections result in the retrograde labeling of only certain areas.

The sculpting of connections during development could involve either the death of cells that project inappropriately (and the survival of cells with normal projections), or the withdrawal of axonal branches. In fact, both cell death and axon withdrawal seem to occur. For example, the elimination of the extensive ipsilateral projections from the eye to the superior colliculus of rodents seems to involve the death of cells that project inappropriately. This has been demonstrated by using retrograde tracers that persist in neurons for long periods of time. Thus, injections of the protein-binding dye fast blue into the superior colliculus of a rat on the day of birth labels retinal ganglion cells throughout the retina ipsilaterally, although there is a higher concentration of labeled cells in the inferior temporal retina that gives rise to the normal projection to the rostromedial portion of the colliculus (Fig. 6.10). If animals are injected on the day of birth (postnatal day 0) and allowed to survive until postnatal day 12, many of the labeled cells outside the inferior temporal retina will have disappeared. An important control involves the removal of the opposite eye; in this case, labeled neurons are still found throughout the retina. This reveals that the disappearance of labeled neurons is not due to the loss of label from cells that survive.

The story in the cerebral cortex is somewhat different, however. Here, retrograde labeling with long-lived markers reveals that the restriction of the cortical projections involves the withdrawal of axons without accompanying cell death. Thus, when cortical neurons that project to the contralateral hemisphere are retrogradely labeled during the time when projections are widespread, these labeled cells can still be found after the ectopic projections have disappeared. Thus, the elimination of ectopic projections can occur either because of the death of cells that project ectopically or because of the withdrawal of the atypically projecting axons.

In addition to projecting to ectopic locations, some axons grow to a point near their target and then stop for a period of time. For example, axons growing into the cerebral cortex reach the level of the white matter some time before they actually enter the cortex. The axons appear to mark time in the white matter, invading the cortex only when the targets are mature enough to begin to accept innervation. The signals that are involved in this sort of "marking time" are not well known, but may involve the formation of transient connections with cells in the white matter.

Although the "ectopic" connections that are formed by some systems during the course of development are eventually eliminated, these should not be thought of as errors of development. Rather, these connections may play an important role in conveying information to the developing neurons, or in regulating either the development of connections or the differentiation of the neurons themselves.

Once axons have reached their target site, the next step involves the formation of the specific synaptic connections. For most neurons this involves a collaboration between presynaptic and postsynaptic partners. In the case of CNS neurons, synapse formation requires that the postsynaptic cell elaborate its receptive surface (the dendritic tree). Thus, the next topic to be considered is the development of the receptive surface of the postsynaptic cell.

Supplemental Reading

General

Jacobson M (1978) *Developmental Neurobiology*. New York, Plenum Press
Purves D, Lichtman JW (1985) *Principles of Neural Development*. Sunderland MA, Sinauer Assoc Inc

The Growth Cone

Bray D (1973) Branching patterns of individual sympathetic neurons in culture. *J Cell Biol* 56:702–712
Bray D (1982) The mechanism of growth cone movements. *Neurosci Res Program Bull* 20:821–829

Speidel CC (1933) Studies of living nerves. II. Activities of ameboid growth cones, sheath cells, and myelin segments, as revealed by prolonged observation of individual nerve fibers in frog tadpoles. *Am J Anat* 52:1–75

Yamada KM, Spooner BS, Wessells NK (1971) Ultrastructure and function of growth cones and axons of cultured nerve cells. *J Cell Biol* 49:614–635

The Precision of Axon Outgrowth

Hibbard E (1965) Orientation and directed growth of Mauthner's cell axons from duplicated vestibular nerve roots. *Exp Neurol* 13:289–301

Lance-Jones C, Landmesser L (1981a) Pathway selection by chick lumbosacral motoneurons during normal development. *Proc R Soc Lond* 214:1–18

Lance-Jones C, Landmesser L (1981b) Pathway selection by chick lumbosacral motoneurons in an experimentally altered environment. *Proc R Soc Lond* 214:19–52

Landmesser (1986) Axonal guidance and the formation of neuronal circuits. *Trends Neurosci* 9/10:489–492

Axon Growth: Contact Guidance

Collins F, Garrett JE (1980) Elongating nerve fibres are guided by a pathway of material released from embryonic non-neuronal cells. *Proc Natl Acad Sci USA* 77:6226–6228

Landmesser L (1986) Axonal guidance and the formation of neuronal circuits. *Trends Neurosci* 9:489–492

The Blueprint Hypothesis

Singer M, Nordlander RH, Egar M (1979) Axonal guidance during embryogenesis and regeneration in the spinal cord of the newt. "The blueprint hypothesis" of neuronal pathway patterning. *J Comp Neurol* 185:1–22

Axon Guidance: Differential Adhesiveness

Gottlieb DI, Rock K, Glaser L (1976) A gradient of adhesive specificity in developing avian retina. *Proc Natl Acad Sci USA* 73:410–414

Halfter W, Claviez M, Schwarz U (1981) Preferential adhesion of tectal membranes to anterior embryonic chick retina neurites. *Nature* 292:67–70

Letourneau PC (1975) Cell-to-substratum adhesion and guidance of axonal elongation. *Dev Biol* 44:77–91

Letourneau PC (1982) Nerve fiber growth and its regulation by extrinsic factors, in Spitzer NC (ed): Neuronal Development. New York, Plenum Press

Rutishauser U, Gall WE, Edelman GM (1978) Adhesion among neural cells of the chick embryo. IV. Role of the cell surface molecule CAM in the formation of neurite bundles in cultures of spinal ganglia. *J Cell Biol* 79:382–393

Axon Guidance: NGF as a Possible Tropic Substance

Campenot RB (1977) Local control of neurite development by nerve growth factor. *Proc Natl Acad Sci USA* 74:4516–4519

Campenot RB (1982a) Development of sympathetic neurons in compartmentalized cultures. I. Local control of neurite growth by nerve growth factor. *Dev Biol* 93:1–12

Campenot RB (1982b) Development of sympathetic neurons in compartmentalized cultures. II. Local control of neurite survival by nerve growth factor. *Dev Biol* 93:13–21

Gunderson RW, Barrett JN (1979) Neuronal chemotaxis: chick dorsal-root axons turn toward high concentrations of nerve growth factor. *Science* 206:1079–1080

Davies AM, Bandtlow C, Heumann R et al. (1987) Timing and site of nerve growth factor synthesis in developing skin in relation to innervation and expression of the receptor. *Nature* 326:353–358

Galvanotropism

Jaffe LF, Poo M-M (1979) Neurites grow faster towards the cathode than the anode in a steady field. *J Exp Zool* 209:115–128

Chemospecificity

Attardi DG, Sperry RW (1963) Preferential selection of central pathways by regenerating optic fibers. *Exp Neurol* 7:46–64

Balsamo J, McDonough J, Lilien J (1976) Retinal-tectal connections in the embryonic chick: evidence for regionally specific cell surface components which mimic the pattern of innervation. *Dev Biol* 49:338–346

Barbera AJ, Marchase RB, Roth S (1973) Adhesive recognition and retinotectal specificity. *Proc Natl Acad Sci USA* 70:2482–2486

Meyer RL, Sperry RW (1976) Retinotectal specificity: Chemoaffinity theory, in Gottlieb G (ed): *Neural and Behavioral Specificity. Studies on the Development of Behavior and the Nervous System* vol 3. New York, Academic Press, pp 111–149

Trisler D, Collins F (1987) Corresponding spatial gradients of TOP molecules in the developing retina and optic tectum. *Science* 237:1208–1209

Guideposts for Axonal Growth

Berlot J, Goodman CS (1984) Guidance of peripheral pioneer neurons in the grasshopper: adhesive hierarchy of epithelial and neuronal surfaces. *Science* 223:493–496

Goodman CS, Ho RK, Ball EE (1982) Pathfinding by growth cones in grasshopper embryos: guidance cues in the ectoderm and mesoderm. *Neurosci Res Program Bull* 20:847–859

Pioneer Axons

Bentley D, Caudy M (1983) Pioneer axons lose directed growth after selective killing of guidepost cells. *Nature* 304:62–65

Bentley D, Keshishian H (1982) Pioneer neurons and pathways in insect appendages. *Trends Neurosci* 5:354–358

The Formation of Neural Connections II: Dendritic Development and the Establishment of Synaptic Connections

The Development of Dendrites

As noted in Chapter 6, the initial outgrowth of both axons and dendrites is apparently determined by factors that are intrinsic to the neuron. Very shortly after emerging from the cell body, axons reorient themselves based upon the tissue environment, responding exclusively to extrinsic influences. A useful generalization is that dendrites are less sensitive to the tissue environment and that many aspects of dendritic form are determined by factors that are intrinsic to the neurons. As will be seen, however, the relative role of intrinsic versus extrinsic influences in the further elaboration of the dendritic arbor is not clear-cut.

Dendrites often develop in parallel with the ingrowth of afferents and the formation of synaptic connections. This is particularly true in cortical areas. For example, in the dentate gyrus of the rat, differentiation of granule cell dendrites begins around birth and is essentially complete by about 30 days of age (Fig. 7.1). Afferent ingrowth and synaptogenesis occur over this same period. A similar pattern of dendritic differentiation occurs in the case of Purkinje cells of the cerebellar cortex; although quantitative comparisons of dendritic development and synaptogenesis have not been carried out, qualitative observations suggest that Purkinje cells also elaborate their dendrites more or less in parallel with the development of parallel fibers.

Studies of other brain regions reveal that the parallel development of dendrites and their afferents is not invariably seen; indeed, some neurons develop elaborate dendritic arbors *before* the arrival of their normal afferents. Studies of the auditory system of the chicken reveal that some cells develop a dendritic arbor that is considerably more extensive than will be maintained in the adult. For example, neurons in the avian homologue of the medial superior olive have bipolar dendrites that receive input from the cochlear nucleus on each side. The length of the dendrites varies systematically throughout the nucleus, so that the portion of the nucleus that receives high-frequency input has the shortest dendrites, and

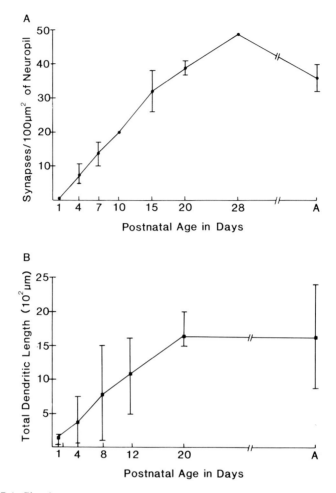

FIGURE 7.1. Simultaneous development of dendrites and their afferents. *A* Quantitative electron microscopic analysis of synapse development in the dentate gyrus of the rat. The graph illustrates the number of synapses per area of neuropil in the molecular layer of the dentate gyrus at various postnatal ages. [From Steward O, Falk PM (1986) Protein synthetic machinery at postsynaptic sites during synaptogenesis: a quantitative study of the association between polyribosomes and developing synapses. *J Neurosci* 6: 412–423. Copyright, Society for Neuroscience.] *B* Quantitative study of the development of dendrites of dentate granule cells. The graph illustrates the mean total dendritic length of individual granule neurons at various postnatal ages. [Redrawn from Cowan WM, Stanfield BB, Kishi K (1980) The development of the dentate gyrus. *Curr Top Dev Biol* 15: 103–157.]

the cells that receive low-frequency input have the longest dendrites. Initially all of the cells in the nucleus develop extensive dendritic arbors. However, the dendritic arbors of cells that come to receive high-frequency sound are substantially pruned during the course of development (Fig. 7.2). A similar overproliferation of dendrites occurs in the avian homologue of the anterior ventral cochlear nucleus. In adult chickens these neurons have large cell bodies and few dendrites. However, early in development the neurons have more extensive dendritic arbors that are withdrawn as the nucleus matures.

These results indicate that dendritic development, like axonal devel-

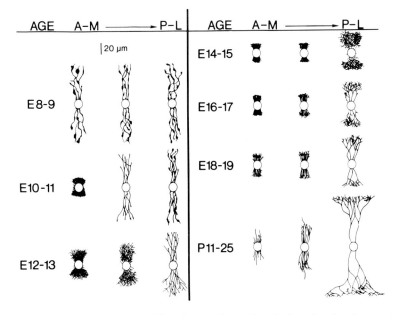

FIGURE 7.2. Dendritic overproliferation and pruning during the development of neurons in brainstem auditory nuclei. Nucleus laminaris (NL) is the avian homologue of the medial superior olive of mammals. Neurons in the nucleus have bipolar dendrites, and the length of the dendrites varies systematically throughout the nucleus. Neurons that receive input from higher frequency regions of the cochlea have the shortest dendrites, whereas neurons that receive low-frequency input have the longest dendrites. Thus, there is a gradient of dendritic length across the anteromedial-posterolateral (AM-PL) axis of the nucleus. During early development, all neurons in the nucleus initially elaborate a more extensive dendritic tree than is maintained at maturity. The figure schematically illustrates representative dendritic arbors of neurons in the various divisions of the nucleus at various ages as revealed by Golgi stains. [From Smith DJ (1981) Organization and development of brain stem auditory nuclei of the chicken: dendritic development in n. laminaris. *J Comp Neurol* 203: 309–333.]

opment, can involve either a monotonic progression from the undifferentiated state to the final form, or an initial stage of overproliferation followed by retraction. The former mode of development is consistent with a role for afferents in regulating dendritic differentiation. The later mode suggests that a dendritic arbor can develop independently of afferents, but that afferents may play a role in determining which dendrites are maintained. Alternatively, it is possible that the correlations between dendritic growth and the formation of synapses by afferents is spurious and that dendritic form is determined entirely by intrinsic factors.

The Role of Afferents in Dendritic Differentiation

To obtain direct evidence on the role of afferents in dendritic differentiation, many studies have evaluated how dendrites develop when afferent innervation is disrupted. The straightforward prediction is that if dendritic development is altered when afferents are disrupted, then this would imply a regulatory role for the afferents. As will be seen, the answer is not simple. Some types of neurons seem to be capable of developing essentially normal dendritic arbors in the absence of afferent innervation or when the normal afferents are severely disrupted. Other cell types seem to be much more sensitive to the presence, and sometimes even the activity, of the afferents.

Perhaps the most dramatic manipulation is to remove neurons entirely from their normal cellular environment by transplanting tissue or by dissociating cells and growing them in tissue culture. These studies have revealed that some aspects of dendritic form can be expressed by neurons in very unusual environments. For example, some cell types can be readily recognized in transplants that are grown in unusual locations in the nervous system; in such locations, these cells certainly could not receive their normal afferents at the appropriate time. In addition, certain characteristic features of dendritic morphology are even expressed by neurons growing in culture, and different types of neurons can be easily distinguished from one another (Fig. 7.3).

The results of experiments that disrupt afferents without removing neurons from their normal in vivo locations also suggest that some neurons can elaborate essentially normal dendritic arbors in the absence of afferent innervation. For example, sympathetic ganglion cells normally elaborate about the same number of dendrites as there are afferent fibers that innervate individual cells (see Fig. 7.18). This relationship obviously implies that there must be some matching between dendrites and their afferents during development; however, dendritic development occurs unimpeded when afferent ingrowth is surgically prevented. Quantitative studies indicate that the neurons continue to elaborate about the same number of dendrites as normal and that these dendrites are of near-normal length.

Although some cells appear to be capable of developing dendrites in

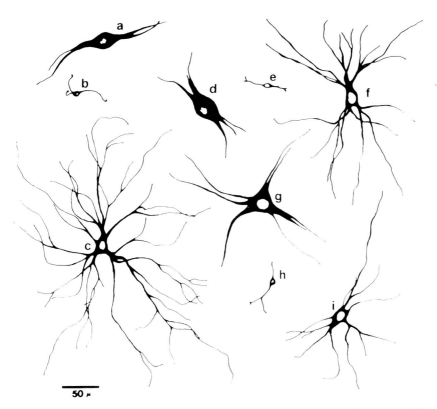

FIGURE 7.3. Dendritic morphology of different types of neurons in culture. The drawings compare the dendritic morphology of sympathetic neurons (*a,d,g*), hippocampal pyramidal neurons (*c,f,i*), and cerebellar granule neurons (*b,e,h*) after 3-6 weeks in culture. Differences in cell body size and in the size and complexity of dendritic branches are evident; generally similar morphologies are expressed by these three cell types *in vivo*. [Data from Fletcher T, and from Banker GA, Waxman AB (1988) Hippocampal neurons generate natural shapes in cell culture, in Lasek RJ, Black MM (eds): Intrinsic determinants of neuronal form and function. *Neurology and Neurobiology*, vol 37. New York, Alan R. Liss, pp 61–82.]

the absence of afferent innervation, for many neurons the final form of the dendrite seems to depend not only on the presence, but also on the functional activity of the synapses that innervate them. Deafferentation of many cell types results in stunted dendrites. For example, in the case of mutations that affect the differentiation of granule cells of the cerebellum, the Purkinje cells fail to receive their normal innervation. The dendrites of these Purkinje cells are much shorter than normal and exhibit less branching (Fig. 7.4). Interestingly, however, the dendrites are still recognizably of Purkinje cell type.

Sensory deprivation can also lead to dramatic alterations in the size and orientation of dendrites in sensory structures. For example, visual deprivation apparently leads to a redistribution of dendrites of stellate cells in the visual cortex. Normally the dendrites of stellate cells arborize extensively within the fourth layer of the cortex, where thalamic afferents terminate; with dark-rearing, the dendrites arborize predominantly in other layers and appear to avoid the fourth layer (Fig. 7.5). This example is interesting because it is the distribution of dendrites that is altered; the dendrites are still easily recognizable as typical of stellate cells. It is not

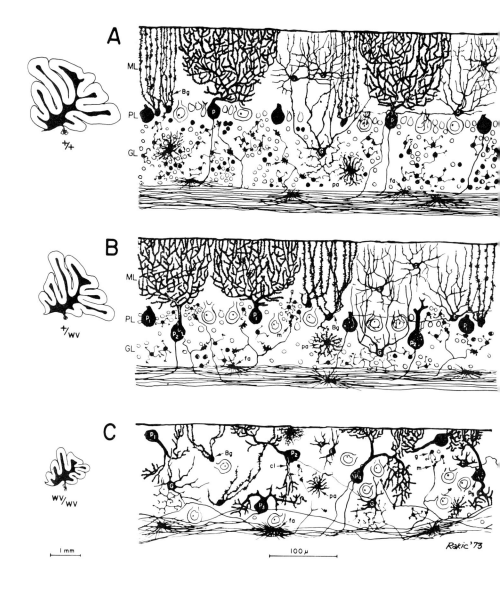

clear whether the average dendritic arbor maintained by individual cells is affected however. This suggests the interesting possibility that neurons may be able to develop dendrites of normal length in the absence of normal innervation if those dendrites can be redirected into areas where they can receive alternative synaptic contacts.

It is important to note that studies of the effects of manipulating afferents are difficult to interpret because it is not clear whether the manipulations affect the *initial growth* of the dendrite or the *maintenance* of the dendrite in its final form once it has developed. This distinction is important. In the former situation, extrinsic influences would be important for the initial elaboration of the dendritic tree. In the latter situation, intrinsic factors would determine the initial form of the dendrite, whereas extrinsic factors would maintain dendrites, preventing their deterioration. One line of evidence in this regard pertains to the effects of deafferentation on fully developed dendrites. Here, the answer is straightforward; deafferentation usually leads to a deterioration of the dendrites, and the extent of the deterioration depends on the extent of the deafferentation (see also Chapter 8). These results clearly indicate that afferents do play a role in maintaining dendrites, but do not reveal whether the afferents also play a role in their initial differentiation.

The preceding discussion reveals that the question of whether dendritic development is influenced by afferents does not have a simple answer. Taken together, the safest conclusion seems to be that the general form of a dendrite is intrinsically determined, in that cells can elaborate dendrites that are characteristic for the cell type even in the absence of their normal afferent innervation. However, the finer aspects of dendritic architecture depend on continuing interactions with afferents. Whether the dependency

◁———————————————————————————————

FIGURE 7.4. Stunting of dendritic growth in the cerebellar cortex of weaver mice. *A* Organization of the cerebellar cortex of normal mice (+ / +). *B* Organization of the cerebellar cortex of heterozygous weaver mice (+ / *wv*). *C* Organization of the cerebellar cortex in homozygous weaver mice (*wv* / *wv*). In homozygous weaver there is a profound loss of granule cells. Thus, the dendrites of Purkinje cells do not receive their normal innervation from parallel fibers. The dendrites of the Purkinje cells form fewer tertiary branches than normal and are misoriented. Interestingly, spines appear and are maintained in the absence of innervation. The interneurons of the molecular layer are even more profoundly affected by the absence of parallel fibers. The dendrites of the interneurons are abortive and randomly oriented. There are few abnormalities in the heterozygous weaver. Bg = Bergmann glia, G = Golgi type II cell, GL = granule layer, ML = molecular layer, PL = Purkinje layer, b = basket cell, cl = climbing fiber, fa = fibrous astrocyte, g = granule cell, m = mossy terminal, p = Purkinje cells, pa = protoplasmic astrocyte, s = stellate cell. [From Rakic P, Sidman RL (1973) Organization of cerebellar cortex secondary to deficit of granule cells in weaver mutant mice *J Comp Neurol* 152: 133–162.]

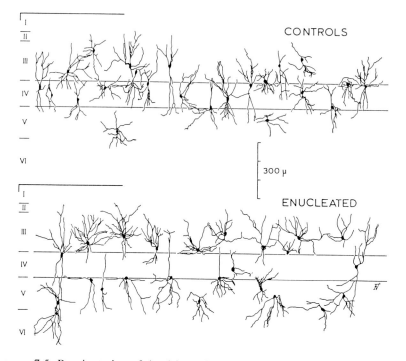

FIGURE 7.5. Reorientation of dendrites of stellate cells of the visual cortex after removal of one eye. The figure illustrates drawings of Golgi-impregnated stellate cells from normal mice (*above*) and mice that had been enucleated at birth. Note that in normal animals, dendrites of stellate cells ramify extensively in layer IV, which is the zone of termination of fibers from the lateral geniculate nucleus. In mice that had been enucleated at birth, the dendrites of stellate cells avoid layer IV, instead ramifying in adjacent laminae. [From Valverde F (1968) Structural changes in the area striata of the mouse after enucleation. *Exp Brain Res* 5: 274–292.]

upon afferents reflects a regulation of dendritic development or dendritic maintenance is difficult to determine.

In sum then, the following general principles seem to apply for the differentiation of axons and dendrites:

1. Axons and dendrites emerge from neuronal cell bodies based on intrinsic factors. The polarity of the neuron is probably determined by the organization of the neuronal cytoskeleton and may be established at the time of the final mitosis.

2. Shortly after leaving the immediate vicinity of the cell body, axons reorient themselves in response to cues in the tissue environment. The guidance of the growing axon then depends on interactions between the growth cone and its cellular environment. Dendrites do not reorient, but continue to grow based on the orientation of the cell body. However,

if these dendrites do not receive afferent connections, they may become stunted and malformed.

3. Axonal and dendritic development can involve either a monotonic progression from the undifferentiated state to the final form, or an initial stage of overproliferation followed by retraction.

4. The final development of dendritic fine structure depends on the presence and functional activity of the afferents to those dendrites. The dependency on afferents is not restricted to the developmental period.

Specificity in the Formation of Topographically Ordered Projections

After growing axons have successfully reached their target areas, a great deal of sorting must still occur as specific connections are formed. This final sorting can best be appreciated by considering the sensory systems. In every sensory system except the olfactory, there is a precise topographic map of the periphery that is maintained at several different levels in the CNS. In the somatosensory system there is a somatotopic order, in the visual system there is a retinotopic order, and in the auditory system there is a tonotopic order. Many other systems exhibit a similar topographic organization, although the relationship to the periphery may be less clear. How such topographically ordered maps are established has been a topic of considerable interest.

There are, in general, two ways that topographically ordered maps could be constructed. The first involves a specificity of axon growth. Specifically, since the topographic order is present in the periphery, axons could simply maintain nearest-neighbor relations throughout their course, establishing connections with target cells on this basis. Studies in the visual system suggest that this interpretation cannot account for the orderly pattern of termination. Although there is some topographic order in the optic nerve, the formation of correct connections does not depend on this order. Thus, in some species the topographic order of axons in the nerve is not precise, but the final connections are. In addition, studies of regeneration of visual projections in lower vertebrates indicate that orderly connections can be formed even when axons are diverted from their normal course. These studies suggest that orderly projections are established as a result of interactions between axons and their targets. Stated another way, topographically ordered projections are established in parallel with the formation of synaptic connections.

The Chemoaffinity Hypothesis

One of the most influential hypotheses about how such topographically ordered connections arise is the *chemoaffinity hypothesis* proposed by Roger Sperry. The research that led to this hypothesis was carried out in the retinotectal system of lower vertebrates. The visual systems of fish and frogs offer special advantages, since they regenerate following injury. Sperry found a selectivity in the pattern of regeneration of retinal projections following damage to the optic nerve. When the optic nerve was cut and the retina was left intact, regenerating fibers reestablished their normal topography in the tectum. However, when the optic nerve was cut and part of the retina was destroyed, the fibers from the surviving portion of the retina regenerated selectively into the parts of the tectum that the fibers would normally innervate (Fig. 7.6). Sperry proposed that selective connections were established because presynaptic and postsynaptic partners have complementary molecular markers that permit recognition and adhesion.

The chemoaffinity hypothesis can be broken down into the following postulates: (1) neurons are intrinsically different from one another, (2) the differences are position-dependent—i.e., properties vary systematically across the topographic axes of the structure in question—, (3) the differences are biochemical in nature and expressed on the external membrane, (4) the differences are present early in development when connections form, and (5) presynaptic and postsynaptic cells with complementary molecular markers connect with each other in a selective and exclusive manner.

Evidence for the first four postulates of the chemoaffinity hypothesis has been obtained by evaluating the aggregation of dissociated cells from the retina to cell aggregates or explants from the tectum (Fig. 7.7). The prediction of the chemoaffinity hypothesis is that cells from different portions of the retina would bind preferentially to cells from different parts of the tectum. This is exactly what was found. For example, cells from the ventral retina adhere preferentially to cells from the dorsal tectum; cells from the dorsal retina adhere preferentially to cells from the ventral tectum. These results indicate that neurons from the different topographic subdivisions are intrinsically different from one another. These differences are position-dependent and parallel the normal topography of the retinotectal projection. Because the differences are revealed in an adhesive assay where cells are intact, the results also suggest that the differences are expressed on the external membrane and may be biochemical in nature. Finally, because the adhesive assays are carried out using dissociated cells obtained from young animals, the results indicate that differences between cells are present early in development.

An enormous amount of effort has been directed toward testing the final postulate of the chemoaffinity hypothesis, that presynaptic and postsynaptic cells with complementary molecular markers connect with each other in a selective and exclusive manner. The hypothesis is attractive for many reasons, since it has great explanatory power. Nevertheless, a variety of experimental tests suggests that the hypothesis is not correct, at least in its most restrictive form. Specifically, the hypothesis predicts

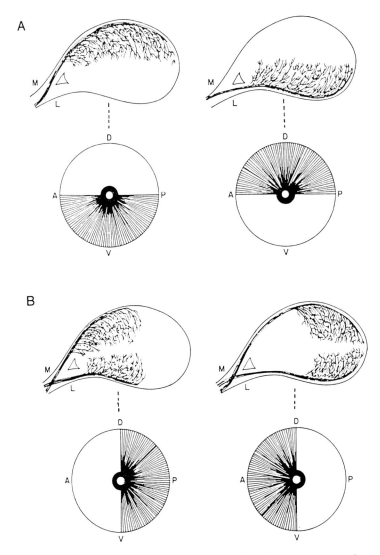

FIGURE 7.6. Specificity in the intial regeneration of optic nerve axons after crush injury in goldfish. The optic nerve was crushed, and a portion of the retina was destroyed so that the pattern of projection of the surviving portion of the retina could be assessed. *A* When the dorsal retina was destroyed, the surviving ventral retina regenerated selectively to the medial portion of the tectum. Similarly, when the ventral retina was destroyed, the dorsal retina regenerated selectively to the lateral portion of the tectum. *B* When the anterior (nasal) retina was destroyed, the surviving posterior retina regenerated selectively to the anterior portion of the tectum. Similarly, when the posterior retina was destroyed, the anterior retina regenerated selectively to the posterior tectum. [From Attardi DG, Sperry RW (1963) Preferential selection of central pathways by regenerating optic fibers. *Exp Neurol* 7: 46–64.]

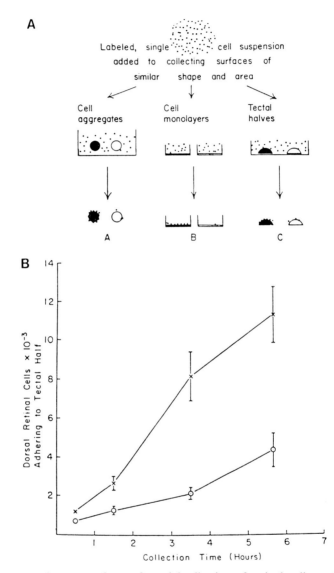

FIGURE 7.7. In vitro assay for preferential adhesion of retinal cells to different parts of the tectum. *A* Single-cell suspensions from the retina were prepared and labeled with a radioactive isotope. The binding of these cells to cells from different parts of the tectum was assessed by evaluating the binding of single cells to cell aggregates, to monolayers of cells grown in culture, or to pieces of tectal tissue. *B* The number of labeled cells from the *dorsal* retina adhering to pieces of ventral (–x–) or dorsal (–O–) tectum. The cells from the dorsal retina adhere preferentially to the ventral tectum, recapitulating the specificity of projection of optic axons upon the tectum. [From Roth S, Marchase RB (1976) An *in vitro* assay for retinotectal specificity, in Barondes SH (ed): *Neuronal Recognition.* New York, Plenum Press.]

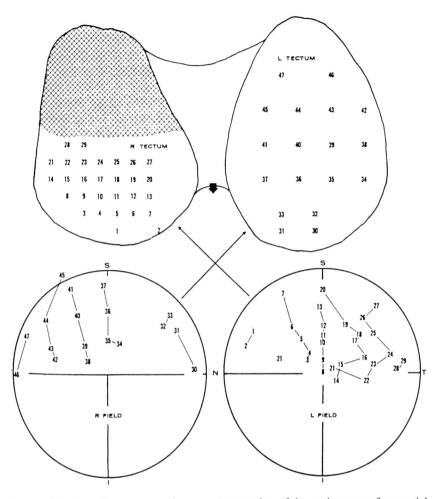

FIGURE 7.8. Size disparity experiments: regeneration of the optic nerve after partial ablation of the tectum. The upper portion of the figure illustrates the two sides of the tectum of a goldfish as they would appear from a dorsal view. The circles below schematically illustrate the visual fields of the fish. Retinotectal connectivity was assessed physiologically by presenting spots of light in different portions of the visual field and recording from the tectum with extracellular electrodes in anesthetized fish. The numbers indicate the position of spot stimuli in the visual field and the location in the tectum where cells respond to a stimulus in the particular portion of the visual field. The normal pattern of retinotectal connectivity is illustrated by the projections from the right eye to the left tectum. The projection from the left retina has regenerated to the half-tectum. When these projections are mapped several months after the destruction of the tectum, the entire visual field is compressed onto the remaining tectum. [From Gaze RM, Sharma SC (1970) Axial differences in the reinnervation of the goldfish optic tectum by regenerating optic nerve fibers. *Exp Brain Res* 10: 171–181.]

that axons of a given presynaptic neuron will connect only with the appropriate target; if this target is not available, then no alternatives should be adequate. This aspect of the hypothesis was tested in a large series of experiments by creating size disparities between the retina and tectum and evaluating the connections that were formed after regeneration had occurred.

Size Disparity Experiments

In one type of *size disparity* experiment, a complete retina was allowed to regenerate back to a partial tectum (one in which half the total target area had been surgically removed). A rigid interpretation of the chemoaffinity hypothesis would predict that the retinal fibers could form connections only with the cells that they would normally innervate; as a result, the portion of the retina that would normally project to the missing portion of the tectum should fail to form connections. Instead, the final retinotectal map was "compressed" in that the entire map was represented on the surviving portion of the tectum. There was a tendency for the regenerating axons to project appropriately during the early stages of regeneration, as had been observed by Sperry; however, over time the connections were adjusted to match the available target cells (Fig. 7.8). The opposite size disparity experiment yielded similar results. For example, when part of the retina was destroyed, there was a tendency for regenerating fibers to project appropriately at first, but eventually the remaining retina regenerated to form a topographically ordered projection to the entire tectum (Fig. 7.9).

The size disparity and other experiments indicate that rigid cell-to-cell specificities probably do not account for the highly ordered topography of the retinotectal system. This conclusion is further supported by studies

—————————————————————————————————▷

FIGURE 7.9. Size disparity experiments: regeneration of retinal fibers after partial destruction of the retina in goldfish. Portions of the retina were ablated, and the optic nerve was crushed. The termination of fibers from the surviving portion of the retina were evaluated electrophysiologically at various times after the crush as described in Fig. 7.8. The convention for illustrating the results is similar to that of Fig. 7.8, except that only the side of the tectum that receives the regenerated input is illustrated. *A* 36 days after retinal ablation and optic nerve crush. At this time the regenerating fibers are restricted to the portion of the tectum that they would normally innervate. *B* 171 days after retinal ablation and optic nerve crush. The retinotectal map has partially expanded so that fibers that would normally terminate only in the rostral tectum extend much more caudally than they normally would. *C* 282 days after retinal ablation and optic nerve crush. By this time, the retinotectal map has expanded so that the surviving portion of the retina distributes fibers throughout the tectum. [From Schmidt JT, Cicerone CM, Easter SS (1978) Expansion of the half retinal projection to the tectum in goldfish: an electrophysiological and anatomical study. *J Comp Neurol* 177: 257–278.]

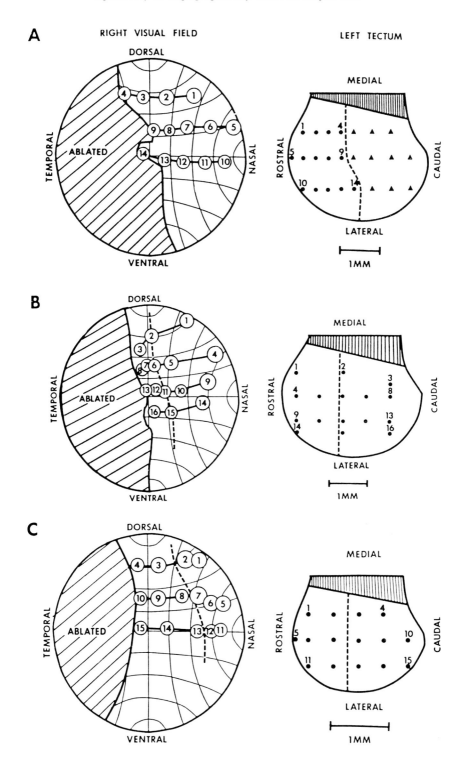

that evaluate the pattern of connectivity during normal development. Studies from a number of different laboratories have revealed that during early development, there is a continuous shifting and reorganization of connections. In fact, such shifting is required because in many species the pattern of cell addition in the retina and tectum is quite different. New cells are added concentrically to the retina, whereas in the tectum, new cells are generated in the caudal tectum. Nevertheless, the connections between retina and tectum are topographically organized throughout the developmental period. The maintenance of topographic order in the face of different patterns of histogenesis would seem to require a reordering of existing connections.

The combined evidence thus suggests that topographic projections are formed on the basis of relative preferences that are then adjusted based upon the situation. The initial establishment of connections may well be based on differences in membrane properties as predicted by the chemoaffinity hypothesis. The later adjustment of the topographic maps seemingly involves interactions between axons and their targets and perhaps between the arbors of individual axons. Because connections are adjusted during the course of development, it can be assumed that key events in the formation of orderly maps are *synapse stabilization* when the connection is appropriate, and *synapse elimination* when contacts are not appropriate.

Rearrangement of Synaptic Connections During Development

The rearrangement of connections on individual postsynaptic cells and the phenomena of synapse stabilization and elimination can be directly demonstrated in certain simple systems where the number of contacts received by postsynaptic cells is precisely regulated. Systems that form one-to-one connections between single axons and single postsynaptic cells offer particular advantages. *In systems where one-to-one connections predominate, single innervation is often preceded by a period of multiple innervation; over time, one of the inputs gains in strength and is stabilized while the other are eliminated.*

Synapse Elimination and Stabilization

In adult mammals, individual muscle fibers are normally contacted by one and only one axon terminal; in contrast, muscle fibers receive innervation from several axons during development. This can be directly demonstrated using physiological techniques. For example, intracellular recordings in muscle fibers of adult rats reveal a postsynaptic potential (psp) of uniform amplitude when graded stimuli are delivered to motor nerves (Fig. 7.10). In developing rats, graded stimulation of motor nerves produces post-synaptic potentials (PSPs) with multiple steps. Multiple steps in PSP am-

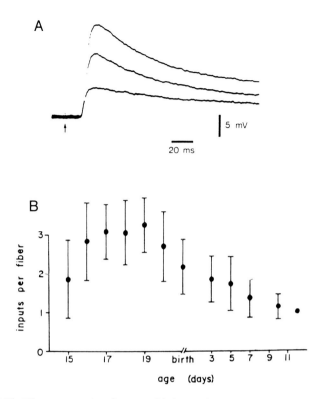

FIGURE 7.10. The progression from multiple to single innervation during the development of neuromuscular connections. *A* The number of inputs to a muscle fiber can be determined by delivering graded stimulation to the motor axon while recording from the muscle. In mature muscles graded stimulation leads to an all or none end plate potential (EPP). In multiply innervated muscles, however, the different axons that project to an individual muscle fiber are usually activated at different stimulus intensities. Thus, graded stimulation produces stepwise increases in EPP amplitude. The number of steps in the EPP indicates the number of different inputs that the muscle receives. *B* Mean number of synaptic inputs per muscle fiber in rats at different developmental ages. Note the progression from multiple to single innervation. [From Dennis MJ, Ziskind-Conhaim L, Harris AF (1981) Development of neuromuscular junctions in rat embryos. *Dev Biol* 81: 266–279.]

plitude indicate that several axons with slightly different thresholds innervate the individual muscle fiber. The number of steps in the PSP indicates the number of different axons innervating the muscle fiber. One can, therefore, plot the average degree of multiple innervation in animals of different ages, as illustrated by Figure 7.10. Thus, muscles are initially innervated by multiple inputs, and during development, all but one input is eliminated.

If individual muscle fibers receive multiple inputs, and some of these are selectively eliminated, it might be expected that individual axons innervate a number of different muscle fibers. Studies of the average size of *motor units* in developing animals reveal that this is indeed the case. Motor unit size is estimated by determining the percentage of total twitch tension developed by a muscle in response to the stimulation of individual axons (in comparison with the tension developed in response to direct stimulation of the muscle). The average size of motor units is often much larger in developing animals and decreases during the first few weeks of life (Fig. 7.11). Although decreases in motor unit size are not invariably observed, they often accompany the process of synapse elimination on individual muscle fibers.

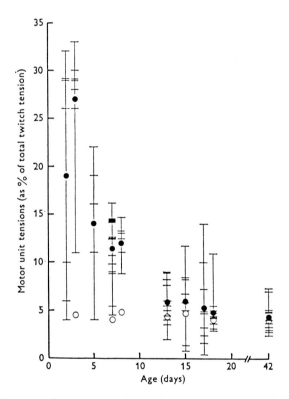

FIGURE 7.11. Decrease in motor unit size during the development of neuromuscular connections in rats. Individual motor axons were stimulated; the percentage of total muscle tension generated by the individual fiber was determined. Whereas multiple innervation (Fig. 7.10) measures the degree of *convergence* of motor axons on individual muscle fibers, measures of motor unit size determine the degree of *divergence* (ie, the number of muscles fibers innervated by individual motor axons). The size of motor units decreases progressively over the first 2 postnatal weeks. [From Brown MC, Jansen JKS, Van Essen D (1976) Polyneuronal innervation of skeletal muscle in new-born rats and its elimination during maturation. *J Physiol* 261: 387–422.]

Similar processes of synapse elimination and synapse stabilization occur during the formation of connections in autonomic ganglia. Intracellular recordings from ganglion cells during graded stimulation of preganglionic fibers reveal that neurons in different ganglia receive a characteristic number of inputs. For example, neurons in the submandibular ganglion of adult rats are typically innervated by one preganglionic fiber; neurons in the rabbit ciliary ganglion receive innervation from two fibers, and neurons in the superior cervical ganglion of the hamster are contacted by about seven axons. In each case the neurons are contacted by a larger number of axons early in development, and the number of contacts is reduced as development proceeds (Fig. 7.12). These results indicate that

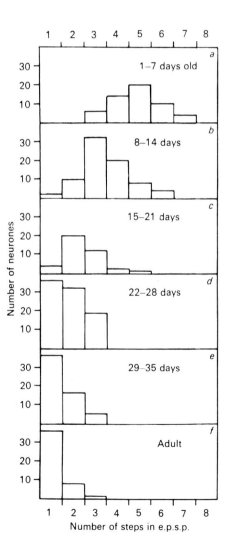

FIGURE 7.12. Reduction in the degree of multiple innervation during the development of synaptic connections in ganglia of the autonomic nervous system. The strategy is the same as that illustrated in Fig. 7.10. Preganglionic fibers to the submandibular ganglion of the rat were activated by stimuli of graded intensity while recording intracellularly from neurons in the ganglion. The number of steps in the EPSP indicates the number of axons innervating the ganglion cells. In mature animals, about 75% of the ganglion cells are innervated by a single axon. This selectivity develops gradually from a pattern of multiple innervation. [From Lichtman JW (1977) The reorganization of synaptic connexions in the rat submandibular ganglion during post-natal development. *J Physiol* 273: 155–177.]

the processes of synapse elimination is not unique to situations in which connections are eventually one-on-one.

Similar types of synapse reorganization have been observed in the CNS. Studies in the avian auditory system have revealed a process of synapse elimination during the establishment of connections between eighth nerve axons and neurons in the avian homologue of the cochlear nucleus (Fig. 7.13). Just after hatching, neurons in the cochlear nucleus receive contacts from an average of four eighth nerve axons. Over the first 2 weeks of life, the number of contacts is reduced, so that neurons in adult chickens are contacted by an average of two eighth nerve axons. A similar process occurs during the establishment of connections between climbing fibers and cerebellar Purkinje cells. Purkinje cells in the cerebellar cortex are normally innervated by one and only one climbing fiber in adult animals. However, these cells are innervated by several climbing fibers in young animals, and the number of climbing fiber inputs to individual Purkinje cells decreases over the first few days of postnatal life (Fig. 7.14).

Rearrangement of Circuitry During the Development of the Cortex

Perhaps the most extensive studies of synapse reorganization in the CNS have been carried out by D. Hubel, T. Wiesel, and their colleagues in the developing visual system. These studies represent part of the work for which Hubel and Wiesel won the Nobel Prize in "medicine or physiology" in 1981. In the visual cortex of higher vertebrates, the projections from the lateral geniculate nucleus are normally segregated into ocular dominance columns—i.e., projections from the laminae in the lateral geniculate nucleus that receive input from contralateral and ipsilateral eyes terminate in alternating columns throughout most of the visual cortex. These columns were initially demonstrated by physiological recording. Thus, if an electrode is advanced tangentially through layer IV of the cortex, one encounters columns of cells that respond predominantly to one eye or the other. The pattern of geniculocortical projections that accounts for the columns can be directly revealed by injecting labeled protein precursors into one eye and allowing the material to be transported transsynaptically into the geniculocortical system related to that eye. Such studies reveal that the terminal fields of geniculocortical fibers from each eye are arranged in alternating stripes (Fig. 7.15).

In developing animals, however, the terminal fields of the projections from the lateral geniculate nucleus laminae related to one eye are uniform, i.e., there is no evidence of stripes (Fig. 7.16). Thus, the projections related to the two eyes must overlap. The overlap of projections predicts that individual cortical neurons might receive inputs from lateral geniculate nucleus fibers related to both eyes; this prediction is amply supported by electrophysiological studies. In fact, before the development of the stripe-like projection pattern, most cortical neurons respond to either eye. The

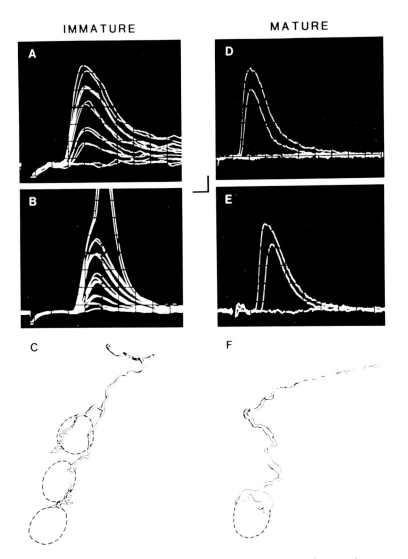

IMMATURE   MATURE

FIGURE 7.13. Reduction in the number of cochlear nerve axons innervating neurons in nucleus magnocellularis (nM) of the chick. The number of inputs to individual neurons was determined by delivering graded stimuli to the cochlear nerve while recording intracellularly from nM neurons (for an explanation of the experimental strategy, see Fig. 7.10). *A, B* Multiple innervation at E13. In both of these examples there are five to six steps in the EPSP with graded stimulation of the cochlear nerve, indicating that five and six axons innervate each nM neuron. *C* Camera lucida drawing of a HRP-filled cochlear nerve axon at E14. Note that this axon branches to innervate three nM neurons. *D, E* In mature chicks there are fewer steps in the EPSP with graded stimulation, indicating fewer axons innervating the nM neurons. *F* HRP-labeling of individual axons from mature chicks indicate the extent of axon branching decreases during the reduction in multiple innervation. [From Jackson H, Parks TN (1982) Functional synapse elimination in the developing avian cochlear nucleus with stimultaneous reduction in cochlear nerve axon branching. *J Neurosci* 12: 1736–1743. Copyright, Society for Neuroscience.]

electrophysiological results are important, since they demonstrate that
the overlapping projections actually form functional synaptic contacts with
neurons in the cortex. Over the course of normal development, the segregated pattern typical of adult animals gradually evolves.

Synapse Competition

There are a number of other examples of initially overlapping projections
that become segregated over the course of development. For example, in
the lateral geniculate nucleus, the projections from the two eyes initially

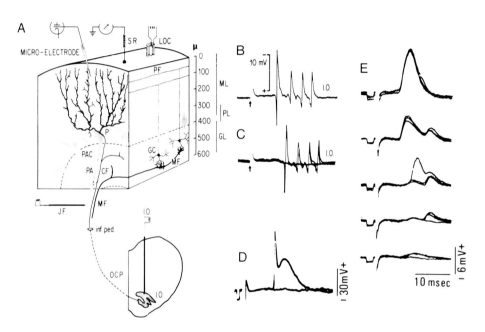

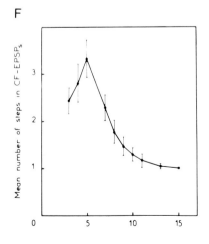

overlap, but then become segregated into nonoverlapping laminae (see Chapter 6). Similar processes may well be involved in the elimination of "exuberant" cortical projections discussed in Chapter 6. In all of these examples the elimination of overlapping projections seems to involve a competition between two sets of very similar axons for populations of postsynaptic cells that either set of axons could potentially innervate. This *competition between synapses* seems to involve the stabilization of one class of synapses and the elimination of others, suggesting an important role for the postsynaptic cell in the selection process.

Synapse competition can be demonstrated directly by evaluating the effects of removing one of the competitors. For example, if muscle groups are deprived of all but a single axon during the time that motor units would normally decrease in size, the motor unit supplied by the surviving axon does not decrease in size. Similarly, the development of segregation of geniculocortical projections is disrupted if one eye is damaged, or even if normal activity is disrupted (e.g., if one eye is deprived of pattern vision by suturing the eyelid). The connections from eyes deprived of normal visual input are less effective in establishing connections with cortical neurons than the afferents from nondeprived eyes. Specifically, after monocular deprivation there is a substantial increase in the width of the ocular dominance columns related to the nondeprived eye at the expense of the columns related to the deprived eye. As a result, most neurons in the cortex respond only to the nondeprived eye (Fig. 7.17).

Synapse competition seems to depend on differential activity in sets of

◁——————————————————————————————

FIGURE 7.14. Progression of multiple to single innervation in the development of climbing fiber (CF) input to Purkinje cells. The strategy is the same as illustrated in Fig. 7.10; stimuli of graded strength were deliverd to the projection while recording intracellularly from Purkinje cells. *A* The experimental arrangement for evaluating CF input to cerebellar Purkinje cells from the inferior olive (IO). *B–D* Extracellular recordings reveal that in adult animals, Purkinje cells discharge in an all-or-none fashion when stimuli of graded strength are delivered to the IO. *B* The response of a Purkinje cell to above-threshold IO stimulation. *C* Two traces when stimulus intensity was just at threshold for the CF. *D* Intracellular recordings illustrating all or none EPSPs in response to a stimulus that was just threshold for the CF from a 10-day-old rat. *E* A series of traces illustrating EPSPs evoked by stimuli of decreasing strength in a 5-day-old rat. Note that there are several steps in the amplitude of the response, indicating the contributions of several CFs. *F* Plot of the average number of steps in CF-EPSPs in rats of different ages. Note the progression from multiple to single innervation [*A–C* from Eccles JC, Ito M, Szentagothai J (1967) *The Cerebellum as a Neuronal Machine.* New York, Springer-Verlag. *D* and *E* from Mariani J, Changeux JH-P (1981) Ontogenesis of olivocerebellar relationships. I. Studies by intracellular recordings of the multiple innervations of Purkinje cells by climbing fibers in the developing rat cerebellum. *J Neurosci* 7: 696–702. Copyright, Society for Neuroscience.]

synapses. During the development of the neuromuscular junction, it is thought that fibers that are more effective in activating the muscle are stabilized, while fibers that are less effective are eliminated. Chronic paralysis decreases the rate of synapse elimination in muscles, whereas chronic stimulation speeds up the elimination of inactive fibers. Exactly the same types of rules apply for the establishment of connections in the visual cortex.

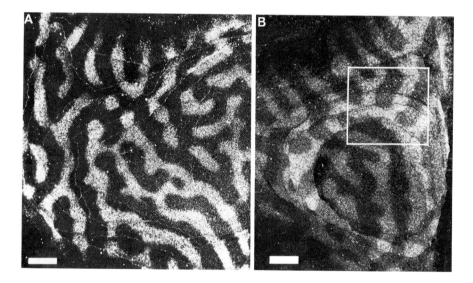

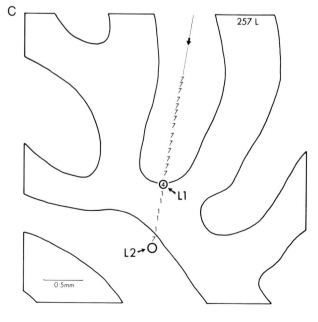

In the cases cited above, competition between synapses leads to a pattern of innervation where one type of afferent "captures" the postsynaptic cell, eliminating other inputs. Interestingly, there is some evidence to suggest that synapse competition may also occur within particular domains on postsynaptic cells. Mature ciliary ganglion cells have a variable number of presynaptic contacts that are highly correlated with the number of individual dendrites. Early in development there is the same variability in the number of dendrites elaborated by individual cells, but all neurons have about the same number of innervating axons. As the animals mature, the "one innervating fiber per one dendrite" rule is established (Fig. 7.18).

There are two important principles that emerge from such observations. First, in establishing the one innervating fiber per one dendrite pattern, given presynaptic fibers seem to capture one dendrite and eliminate other inputs. These results suggest that competition may operate on a dendrite-by-dendrite basis. In addition, *the form of the postsynaptic cell (the number of primary dendrites) is correlated with the number of presynaptic inputs it will eventually receive.*

In the CNS somewhat different rules apply. Generally, neurons with no dendrites or with simple dendritic arbors are innervated by synapses derived from one or a few axons, whereas neurons with complex dendritic arbors receive a number of different inputs. Indeed, the number of different inputs is usually related to the complexity of the dendritic arbor. For many neurons, different types of afferent systems terminate on different portions

◁───

FIGURE 7.15. Ocular dominance columns in the visual cortex of the monkey. Ocular dominance columns in layer IV of the primate visual cortex have been documented in two ways. Anatomically, the projections from the lateral geniculate nucleus (LGN) that relay information from one eye can be traced by injecting 3H-proline into one eye and allowing sufficient time for transport to the terminals of retinal ganglion cells, transsynaptic transfer of the label to LGN neurons innervated by these retinal fibers, and transport of the labeled proteins to the terminals of the LGN neurons in the cortex. *A* Autoradiograph prepared from a tangential section through layer IV of the visual cortex ipsilateral to the injected eye in a 6-week-old monkey. The white stripes indicate the terminal fields of labeled LGN fibers. Note that labeled terminal fields alternate with unlabeled stripes. The unlabeled stripes contain the terminal fields of LGN fibers associated with the contralateral eye. *B* Autoradiograph prepared from a tangential section through the visual cortex contralateral to the injected eye. The box indicates the area enlarged in *C*. *C* Path of a recording electrode through layer IV. The numbers refer to the ocular dominance of the neurons that were encountered. The number 7 indicates a response to the ipsilateral eye only; 1 = a response to the contralateral eye only; 4 = an equivalent response to either eye. L2 = the location of a small marking lesion. [From LeVay S, Wiesel TN, Hubel DH (1980) The development of ocular dominance columns in normal and visually deprived monkeys. *J Comp Neurol* 191: 1–51.]

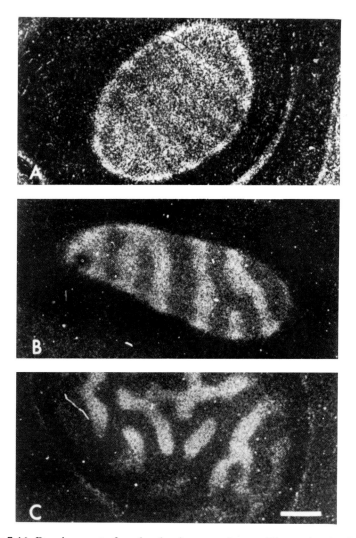

FIGURE 7.16. Development of ocular dominance columns. The ocular dominance columns were traced autoradiographically as described in Fig. 7.15. Each panel illustrates a tangential section through layer IV of the visual cortex. *A* 1 week postnatal; *B* 3 weeks postnatal; *C* 6 weeks postnatal. In the 1-week-old animal, there is little evidence for ocular dominance columns. Instead, the LGN axons from one eye appear to terminate throughout layer IV. By 3 weeks a columnar pattern of termination is evident, although the pattern is not as sharply defined as in the adult. By 6 weeks the pattern resembles the adult. [From LeVay S, Wiesel TN, Hubel DH (1981) The postnatal development and plasticity of ocular dominance columns in the monkey, in Schmitt FO, Worden FG, Adelman G, et al. (eds): *The Organization of the Cerebral Cortex.* Cambridge, MIT Press.]

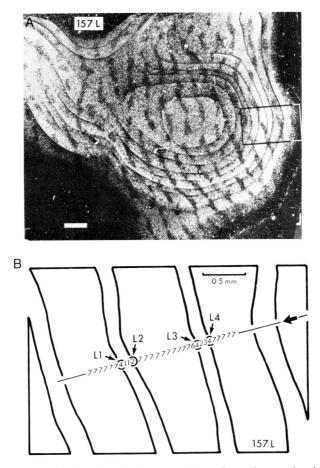

FIGURE 7.17. Expanded ocular dominance column from the nondeprived eye in a monkey whose right eye was closed by eyelid suture at 2 weeks of age. *A* Ocular dominance columns from the nondeprived eye were traced autoradiographically 18 months later as described in Fig. 7.15. Note that the columns from the nondeprived eye are much wider than normal. The unlabeled bands (which would contain the terminals of LGN axons associated with the deprived eye) are much narrower than normal. The box indicates the area schematized in *B*. *B* Path of a recording electrode through layer IV. The numbers refer to the ocular dominance of the neurons that were encountered. The number 7 indicates a response to the ipsilateral eye only; 1 = a response to the contralateral eye only; 4 = an equivalent response to either eye. [From LeVay S, Wiesel TN, Hubel DH (1980) The development of ocular dominance columns in normal and visually deprived monkeys. *J Comp Neurol* 191: 1–51.]

of the dendrites. For example, one type of pathway may terminate on proximal dendrites, and another pathway may terminate on distal dendrites. Thus, for CNS neurons, competition between different afferent systems may be prevented by creating a geometry that leads to the separation of afferent systems that might otherwise compete with one another for sites on the postsynaptic cell.

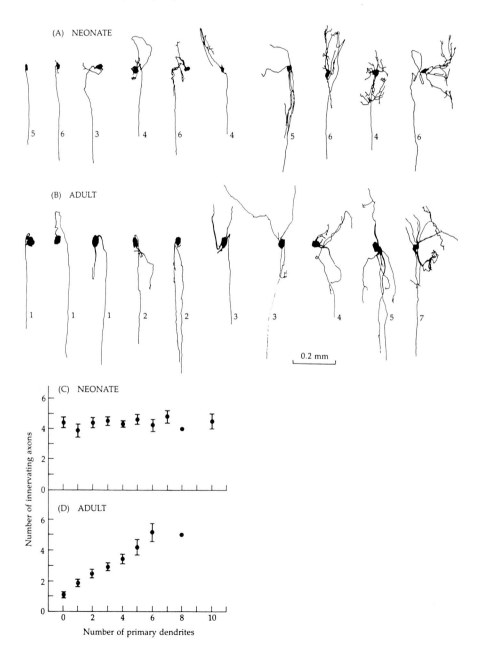

Construction of Synaptic Connections

The construction of synaptic connections requires that the molecular constituents be delivered to the synaptic site; membrane components of the synapse must be inserted into presynaptic and postsynaptic membranes, and cytoplasmic components of the synapse must somehow be positioned at the appropriate location. In the presynaptic membrane this involves the accumulation of vesicles and the assembly of the machinery for stimulus-coupled release. In the postsynaptic membrane the molecules of the postsynaptic membrane specialization must be assembled. These include receptors and their associated ion channels, recognition and adhesion molecules, enzymes such as protein kinases and phosphodiesterases, and a variety of structural molecules. Relatively little is known about the precise cellular and molecular events that occur during the construction of synaptic connections. Most of what is known involves the assembly of patches of ACh receptors beneath synaptic connections that are formed between spinal cord neurons and muscle cells grown in culture.

Studies by G. Fischbach and his colleagues have taken advantage of the fact that it is possible to map the distribution of ACh receptors over time using physiological methods that could be subsequently confirmed anatomically with the use of selective labels (i.e., α-bungarotoxin, see Fig. 7.19). These studies revealed that prior to innervation, ACh receptors are diffusely distributed over the muscle surface, although there are a few "hot spots" with higher receptor density (Fig. 7.19). At mature neuromuscular junctions, however, ACh receptors are most abundant in the

◁―――――――――――――――――――――――――――――――――――

FIGURE 7.18. Synaptic rearrangement during the development of input to ciliary ganglion cells. Ciliary ganglion cells have a variable number of dendrites, and in mature animals there is a close correlation between the number of dendrites and the number of presynaptic inputs that a given cell receives. This has been established by combining physiological analyses of the number of inputs to individual neurons (see Fig. 7.10) with anatomical methods by injecting the physiologically characterized cells with horseradish peroxidase (HRP), which fills the cells and permits an analysis of dendritic number. A, B Examples of drawings of HRP-filled neurons in neonatal (A) and adult (B) animals. Numbers indicate the number of dendrites. C, D Relationship between the number of inputs that a given cell receives and the number of dendrites on that cell. In developing animals the variability in dendrite number is comparable to that of the adult, but there is no correlation between the number of dendrites and the number of inputs that a given neuron receives. [From Purves D, Lichtman JW (1985) *Principles of Neural Development*. Sunderland. MA, Sinauer Assoc. After Purves D, Hume RI (1981) The relation of postsynaptic geometry to the number of presynaptic axons that innervate autonomic ganglion cells. *J Neurosci* 1: 441–452, Copyright, Society for Neuroscience; and Hume RI, Purves D (1981) Geometry of neonatal neurones and the regulation of synapse elimination. *Nature* 293: 469–471. Reprinted by permission of *Nature* 293: 469–471. Copyright © 1981 Macmillan Magazines Ltd.]

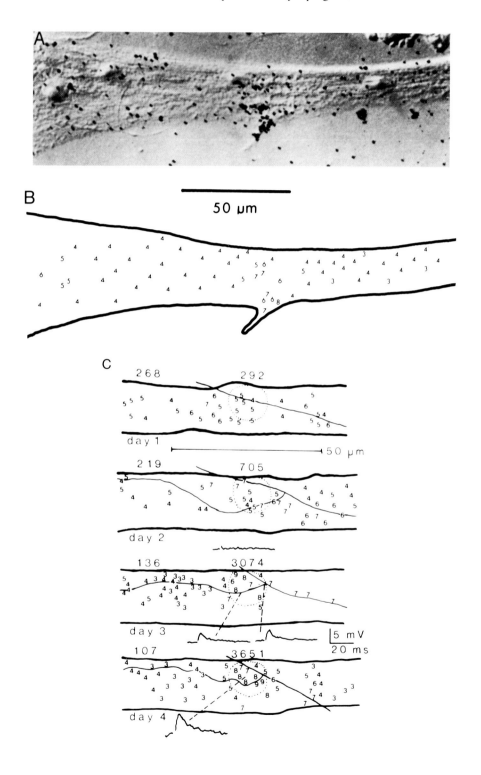

area just beneath the synaptic site. It was initially thought that axons might preferentially innervate the areas of high receptor density; however, this proved not to be the case. Instead, when growing axons contacted the muscle fibers, they formed synapses in a pattern that bore no obvious relation to the receptor distribution prior to innervation. As the synapses formed, new hot spots were established beneath the synaptic sites, so that the diffuse pattern of receptor distribution was transformed into a pattern where receptor density was highest just beneath the synapse.

Logically, there are two ways that a diffuse pattern of receptor distribution could become restricted: (1) receptors that are already present in the membrane could accumulate beneath the synapse; (2) new receptors could be added beneath the synaptic site, with "extra-junctional" receptors being eliminated. In fact, receptor patches beneath synapses seem to be constructed mostly from newly synthesized receptors, although some receptors that were present before the establishment of the patch also contribute. Because the ACh patches beneath synapses are constructed from newly synthesized receptors, most of the extrajunctional receptors present before the establishment of innervation are presumably eliminated, perhaps as a result of endocytosis and degradation by the muscle.

If the receptor patches beneath synapses are constructed mostly from newly synthesized receptors, two important questions are (1) how is receptor synthesis regulated?, and (2) how are the receptors targeted to the appropriate region on the muscle membrane? As noted in Chapter 4, ACh receptors may be synthesized within cytoplasmic domains just beneath

◁————————————————————————————————————

FIGURE 7.19. The establishment of ACh receptor clusters during the formation of neuromuscular connections. These experiments evaluated cultures of myotubes during the establishment of innervation by neurites from spinal cord explants. The distribution of ACh receptors can be mapped anatomically by evaluating the distribution of α-bungarotoxin binding sites (A), and physiologically by measuring ACh sensitivity at different locations on the myotube (B). For the physiological studies, a recording electrode was positioned intracellularly in the myotube, and ACh was applied iontophoretically by a separate electrode. The scale for the ACh sensitivity is (1) 10-22 mV/nC, (2) 22-40 mV/nC, (3) 40-100 mV/nC, (4) 100-220 mV/nC, (5) 220-460 mV/nC, (6) 460-1,000 mV/nC, (7) 1,000-2,200 mV/nC, (8) 2,2000-4,600 mV/nC (9) 4,600-10,000 mV/nC. C the physiological measurements can be made repeatedly over a period of days. Thus, it is possible to monitor ACh sensitivity during the establishment of synaptic connections. The sequence in C illustrates ACh sensitivity over a single myotube over a 4-day period, during which an axon established a synaptic contact in the encircled area. On day 1, ACh sensitivity within the circle was similar to the background; by day 2 the sensitivity within the circle was beginning to increase and continued to increase on each subsequent day. [Reproduced from *The Journal of Cell Biology* 1979; 83: 143–158 by copyright permission of the Rockefeller University Press.]

the synaptic site since the mRNA for the receptors is localized there. Very little is known about how the other molecular constituents of the synaptic region are synthesized, delivered to synaptic sites, and assembled. Clearly, the mechanisms will be considerably more complex in neurons that receive thousands of synaptic connections of many different types.

One interesting possibility is that CNS neurons synthesize some of the constituents of the synapse "on-site" rather than depend on a transport machinery to deliver recently synthesized protein constituents to synapses. The polyribosomes that are selectively positioned beneath synaptic sites on CNS neurons provide a potential cellular mechanism for such local synthesis. This protein synthetic machinery is particularly prominent beneath developing synapses (Fig. 7.20). The issue of how the many postsynaptic sites on neurons are constructed and maintained will be an important topic of research for years to come.

Transsynaptic Regulation of Neuronal Differentiation

In the establishment of connections between neurons or between neurons and other cells, there are many examples where the connections themselves seem to contribute to further differentiation of the interconnected partners. Transsynaptic regulation can be either retrograde or orthograde. For example, as discussed in Chapter 5, naturally occurring cell death during normal development is regulated partly through the establishment

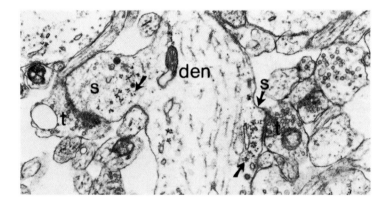

FIGURE 7.20. Accumulations of polyribosomes under developing synapses on neurons in the CNS. A developing spine synapse from the dentate gyrus of the rat is illustrated. s = spine; t = terminal; *arrows* indicate polyribosomes. [From Steward O, Falk PM (1985) Polyribosomes under developing spine synapses: Growth specializations of dendrites at sites of synaptogenesis. *J Neurosci Res* 13: 75–88.]

of connections. Cell death can be enhanced by eliminating targets and prevented if additional target tissue is provided, or if competition is eliminated. Regulation of the differentiation of one neuron by its target is termed *retrograde transsynaptic regulation*. Similarly, some neurons do not survive beyond a certain period if they are deprived of their normal innervation. This represents a form of *orthograde transsynaptic regulation*.

Transsynaptic regulation is important not only for determining the survival of the interconnected cells, but also for contributing to differentiation. One of the better understood examples of these transsynaptic regulatory effects involves the role of innervation in establishing muscle fiber type.

Neural Control of Muscle Fiber Type

There are two general types of striated muscles, which differ in their speed of contraction (so-called fast and slow muscles). The speed of contraction is determined by the type of myosin produced by the muscle cell; thus, muscle type depends on the molecular constituents of the muscle. Fast and slow muscle fibers can be intermingled in an individual muscle, but all of the fibers innervated by a single motor neuron are of the same type. Thus, either motor axons grow specifically to a given type of muscle fiber, or the axons actually determine the muscle fiber type after contact.

In a classic experiment, Buller et al. (1960) cut the axons to two adjacent fast and slow muscles, and allowed the axons to regenerate to the inappropriate muscle (cross-innervating). They found that fast and slow muscles were interconverted as a result of cross-innervation (Fig. 7.21). Subsequent studies have revealed that the changes in physiological properties are correlated with changes in myosin type, suggesting a change in gene expression. The conversion of muscle types can also be brought about by artificial stimulation with patterns of activity typical of "fast" versus "slow" axons. Again, the changes in contractile properties are correlated with changes in the type of myosin produced by the muscle. Thus, the pattern of activity generated by afferents can regulate gene expression by the postsynaptic cell. Recent evidence suggests that afferent regulation may not be the entire story behind the matching between afferents and muscle type, since uninnervated myotubes also express different myosin types. Nevertheless, the demonstration that muscles can be converted indicates that afferents can play an important role in regulating gene expression.

Other evidence that afferents can influence postsynaptic differentiation comes from studies of the development of sympathetic ganglia. In these ganglia, the production of tyrosine hydroxylase (the rate-limiting enzyme in norepinephrine synthesis) depends on afferent innervation. When preganglionic fibers are disrupted, the increases in production of tyrosine hydroxylase, which normally occur during development, are not observed

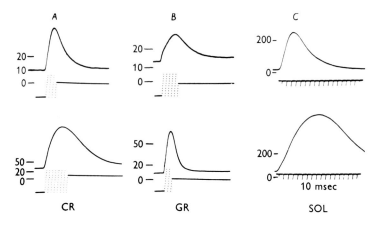

FIGURE 7.21. Afferent regulation of muscle fiber type. The contractile properties of muscle can be altered by modifying their innervation. *A* When a normal slow muscle (crureus) is reinnervated by axons that normally innervate a fast muscle (gracilis), the contraction speed of the crureus increases. The upper trace illustrates the contraction speed of the reinnervated crureus muscle; the lower trace illustrates the contraction speed of the normal crureus (the control muscle from the opposite side). *B* Similarly, when a muscle that is normally fast is reinnervated by a nerve that normally supplies a slow muscle, the contraction speed of the normally fast muscle decreases. The upper trace illustrates the contraction speed of the reinnervated gracilis muscle; the lower trace illustrates the contraction speed of the normal gracilis muscle. *C* Another example of the conversion of a fast to a slow muscle. The upper trace illustrates the contraction speed of the soleus muscle after cross-union with nerves to the peroneal muscle. [From Buller AJ, Eccles JC, Eccles RM (1960) Interactions between motoneurones and muscle in respect of the characteristic speeds of their responses. *J Physiol* 150: 417–439.]

(Fig. 7.22). Thus, afferent activity regulates the production of the enzymes that determine the postsynaptic cell's capabilities to communicate with other neurons.

Transsynaptic regulatory effects are probably quite common in normal development. Such effects are conceptually identical to induction via cell-cell interactions that occur during the development of the nervous system and in other tissues. The only difference is that the transsynaptic effects are cell-cell interactions mediated by a synapse. Of considerable interest are the mechanisms of these regulatory processes. Are they a result of physiological activity and thus perhaps a means by which cells can alter their connections based on functional effectiveness, or are they mediated by some other aspect of the contact? Most of the specific experiments designed to investigate these interactions involve studies of the effects of the elimination of the connection (either depriving a presynaptic axon of

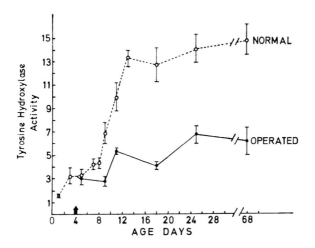

FIGURE 7.22. Transsynaptic regulation of neuronal differentiation. The effects of denervation on the development of neurotransmitter synthesis by sympathetic ganglion cells. Tyrosine hydroxylase is the rate-limiting enzyme for the synthesis of norepinephrine in ganglion cells of the sympathetic nervous system. Normally, the level of tyrosine hydroxylase activity increases in sympathetic neurons over the postnatal period. When the ganglion cells are denervated by cutting the cervical sympathetic trunk, the levels of tyrosine hydroxylase activity do not increase. *Arrow*-the time of denervation. [From Black IB, Hendry IA, Iversen LL (1972) Effects of surgical decentralization and nerve growth factor on the maturation of adrenergic neurons in a mouse sympathetic ganglion. *J Neurochem* 19: 1367–1377.]

its target or eliminating inputs to a postsynaptic cell). These studies reveal synapse-mediated interactions that are important throughout the life of the organism and are typically considered under the topic of "trophic influences," which will be discussed further in Chapter 8.

Supplemental Reading

Elaboration of Dendrites

Banker GA, Cowan WM (1979) Further observations on hippocampal neurons in dispersed cell culture. *J Comp Neurol* 187:469–494

Van der Loos H (1965) The "improperly" oriented pyramidal cell in the cerebral cortex, and its possible bearing on problems of neuronal growth and cell orientation. *Bull Johns Hopkins Hosp* 117:228–250

Synaptogenesis

Bloom FE (1972) The formation of synaptic junctions in developing rat brain, in Pappas GD, Purpura DP (eds): *Structure and Function of Synapses*. New York, Raven Press, pp 101–120

Rees RP, Bunge MB, Bunge RP (1976) Morphological changes in the neuritic growth cone and target neurons during synaptic junction development in culture. *J Cell Biol* 68:240–263

Synapse Stabilization and Elimination

Brown MC, Jansen JKS, Van Essen D (1976) Polyneuronal innervation of skeletal muscle in new-born rats and its elimination during maturation. *J Physiol* 261:387–422

Changeux J-P, Danchin A (1976) Selective stabilization of developing synapses as a mechanism for the specification of neuronal networks. *Nature* 264:705–712

Hume RI, Purves D (1981) Geometry of neonatal neurones and the regulation of synapse elimination. *Nature* 293:469–471

Jackson H, Parks TN (1982) Functional synapse elimination in the developing avian cochlear nucleus with simultaneous reduction in cochlear nerve axon branching. *J Neurosci* 2:1736–1743

Landmesser L, Pilar G (1974) Synaptic transmission and cell death during normal ganglionic development. *J Physiol* 241:737–749

Lichtman JW (1977) The reorganization of synaptic connexions in the rat submandibular ganglion during post-natal development. *J Physiol* 273:155–177

Competition During Synaptogenesis in the Cortex

Hubel DH, Wiesel TN (1970) The period of susceptibility to the physiological effects of unilateral eye closure in kittens. *J Physiol* 206:419–436

Wiesel TN (1982) Postnatal development of the visual cortex and the influence of environment. *Nature* 299:583–591

Wiesel TN, Hubel DH (1963) Single cell responses in striate cortex of kittens deprived of vision in one eye. *J Neurophysiol* 26:978–993

Elimination of Exuberant Projections in the CNS

Innocenti GM (1981) Growth and reshaping of axons in the establishment of visual callosal connections. *Science* 212:824–827

Ivy G, Killackey HP (1982) Ontogenetic changes in the projections of neocortical neurons. *J Neurosci* 2:735–743

Rakic P, Riley KP (1983) Overproduction and elimination of retinal axons in fetal rhesus monkey. *Science* 219:1441–1444

Stanfield BB, O'Leary DDM, Cowan WM (1982) Selective collateral elimination in early postnatal development restricts cortical distribution of rat pyramidal tract neurones. *Nature* 298:371–373

Accumulation of ACh Receptors at the Neuromuscular Junction

Frank E, Fischbach GD (1979) Early events in neuromuscular junction formation *in vitro:* induction of acetylcholine receptor clusters in the postsynaptic membrane and morphology of newly formed synapses. *J Cell Biol* 83:143–158

Transneuronal Regulation of Postsynaptic Differentiation

Black IB (1978) Regulation of autonomic development. *Ann Rev Neurosci* 1:183–214

Close R (1965) Effects of cross-union of motor nerves to fast and slow skeletal muscles. *Nature* 206:831–832

Guth L, Watson PK (1967) The influence of innervation on the soluble proteins of slow and fast muscles of the rat. *Exp Neurol* 17:107–117

Salmons S, Vrbova G (1969) The influence of activity on some contractile characteristics of mammalian fast and slow muscles. *J Physiol* 201:535–549

Walicke PA, Campenot RB, Patterson PH (1977) Determination of transmitter function by neuronal activity. *Proc Natl Acad Sci USA* 74:5767–5771

CHAPTER 8

Nervous System Regeneration and Repair Following Injury

Introduction

Injuries and disease that result in neuronal loss or degeneration represent a substantial health problem. Spinal cord injuries leading to paraplegia or quadraplegia affect about 100,000 people, head injuries about 3 million, stroke about 2 million, and degenerative diseases of various sorts (Alzheimer's disease, Parkinson's disease, and Huntington's disease) affect millions more. Thus, problems that lead to the degeneration of CNS neurons are quite common.

It is of particular importance that injuries to the CNS occur very frequently in the young. Brain damage at birth is a common cause of disability. In addition, brain and spinal cord injuries occur most frequently in that segment of the population that is most physically active. Automobile accidents account for a substantial proportion of these, and other common causes are sports activities. For the most part, the prognosis for recovery from CNS injury or disease is not good, yet as a result of vastly improved health care, affected patients can expect a nearly normal life span. Since injuries are most common in the young, patients are affected for many years, often for most of their lives. Obviously, because CNS injuries are so debilitating, the quality of life is a major concern.

Until very recently there were few direct treatments for CNS-injured patients. Most of the efforts were directed toward stabilizing the patients' physiological condition, preventing secondary health problems (e.g., bladder infections, circulatory disorders, etc.), and rehabilitating with training in alternative strategies. During the last decade, however, basic research has provided sufficient understanding so that specific treatment strategies are being developed with the aim of preventing or reversing the damage brought about by injury or disease. Some of the emerging treatment strategies involve (1) preventing the cascading deterioration that results from the loss of normal trophic influences, (2) stimulating and promoting the limited growth responses that normally occur following injury, (3) attempting to induce growth that would otherwise not occur, (4) replacing tissue lost to injury or disease with transplants, and (5) developing prostheses, particularly for sensory systems where the peripheral receptors or the connections between the periphery and the CNS have been disrupted.

All of these new approaches have come about because of advances in understanding the basic cellular and molecular properties of CNS neurons and the responses of the nervous system to damage. This chapter summarizes some of the recent advances and indicates where these advances have already found clinical application.

The Nature of a Lesion from a Cellular Perspective

Before considering how the nervous system adapts to injury, it is important to consider what damage to the nervous system actually involves. In the case of actual physical trauma in mature organisms, the consequences for neuronal circuitry include:

1. A loss of neurons. Since neurons are not capable of proliferating after the developmental period, there is no *cellular regeneration;* therefore, neurons lost to trauma are not replaced. This means not only that the projections of these neurons to their normal targets are lost, but also that the targets of other neurons are removed (Fig. 8.1). *Unless trauma is restricted to a fiber system (peripheral nerve or fiber tract), simple regeneration of damaged axons cannot restore damaged circuitry.*
2. An interruption of axons passing through the damaged area. Axonal interruption also usually leads to *deafferention* or *denervation* (the loss of normal inputs to a neuronal population) and the interruption of connections between axons and their normal targets. For the presynaptic axon, interruption of the axon is functionally equivalent to *target loss* (Fig. 8.1). Damage to axons also leads to a loss of a portion of the neuronal cytoplasm. By analogy with trees, this is termed *axonal pruning.*

In addition to these direct effects on neuronal circuitry, trauma also involves destruction of the glial cells that normally reside in the damaged region, disruption of the blood-brain barrier (which may have important toxicological and immunological consequences), and, perhaps, a disruption of blood supply to some other sites.

In developing animals some of the same effects also occur. However, depending on the age at the time of the trama, other effects may predominate. For example, damage along the route of a growing fiber pathway may prevent the formation of connections rather than disrupting connections, which already exist. In this case, the "presynaptic" neurons would be deprived of their normal targets and the "postsynaptic" neurons of their normal inputs, but there may be no direct physical damage to either pre- or postsynaptic cells. Thus, there would be *deafferentation* and *target*

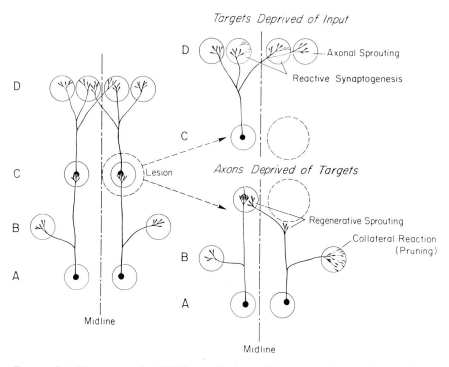

FIGURE 8.1. The nature of a CNS lesion from a cellular perspective. A hypothetical series of pathways with 4 "stations" (A–D) is illustrated on the left. These pathways exist bilaterally and have crossing components at station D. With a unilateral lesion at station C, neurons from station A lose their targets. If the lesion occurs when the pathways are fully developed, the axons of the neurons in station A are amputated. Some cells at station D are denervated. In addition, the cells of origin of some of the projections to station D are lost. The right-hand side of the figure illustrates some of the growth responses that can occur during the reorganization of circuitry following lesions. [From Steward O (1982) Assessing the functional significance of lesion-induced neuronal plasticity. *Int Rev Neurobiol* 23: 197–253.]

loss, but no *pruning.* Trauma may also alter cell-cell interactions that are necessary for normal differentiation, and may also remove precursor cells for parts of the nervous system that had not yet developed.

Under some circumstances, a lesion may be more restricted in terms of its cellular consequences. For example, damage to well-defined nerve trunks may lead to deafferentation and target loss, but not to direct loss of neurons. In addition, the loss of neurons in some disease states may be rather selective. The loss of neurons in Parkinson's disease, for example, may occur in the absence of direct axonal damage to other systems. In this case the effects would be limited to *deafferentation* and *target loss.*

The same may be true of other degenerative diseases. For example, there is evidence for a selective loss of some neuronal populations in Alzheimer's disease; this may occur without direct damage to the axons of other neurons.

In evaluating the reactive responses of the nervous system to injury, it is important to keep in mind these aspects of the lesion. The different reactive events that are set into motion by damage may, in fact, have different root causes. Thus, differences in the nature of the damage (from the cellular perspective) may determine which reactive processes are set into motion.

Regressive Phenomena Following Injury

Consequences of Injury to Partially Damaged Neurons

When damage occurs directly to the neuronal cell body, the entire neuron dies. This is true because both axons and dendrites depend on the cell body for the production of proteins. Similarly, when axons or dendrites are separated from their cell bodies, they cannot survive. When the damage is restricted to the axonal or dendritic compartment and the cell body is spared, the result may be the death of the neuron or survival in a modified state.

Neurons that are *axotomized* often exhibit an atrophy or even complete death. This is termed *retrograde degeneration* (Fig. 8.2). The eventual death of the neuron could be a result of the actual physical damage or the interruption of trophic support from a target (see below).

Evidence that retrograde degeneration is related to actual damage to the axon is that the proportion of cells that dies is often a function of how much axoplasm is lost. Thus, axotomy of motor neurons leads to greater retrograde degeneration when the damage occurs close to the cell body. The extent of axoplasmic loss may not be the major determinant, however, since retrograde degeneration also depends on whether the neuron retains any connections. If axotomized neurons have collateral projections that are not damaged, such neurons often do not exhibit retrograde degeneration even when a substantial portion of their axoplasm is lost. The hypothesis that such collaterals sustain neurons after axotomy is known as *the concept of sustaining collaterals* (Fig. 8.2B).

There are relatively few studies on the effects of destroying dendrites *(dendrotomy)*. Nevertheless, these sorts of injuries must be quite common, particularly in the cerebral cortex. In many cortical injuries the actual physical trauma may be limited to superficial layers. These contain the apical dendrites of pyramidal cells in deeper layers. The extent to which such *dendrotomy* affects cell survival is unclear.

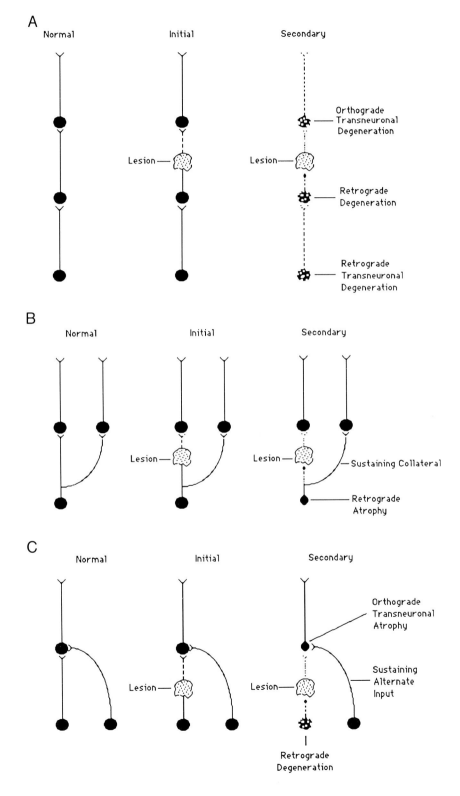

Trophic Phenomena

The loss of normal inputs or targets can lead to atrophy or death of the affected cells. These effects are termed *transsynaptic*, since the affected neurons are synaptically related to the system that are directly damaged. When cells die as a consequence of the removal of their targets, the phenomenon is termed *retrograde transneuronal degeneration*, and when neurons die following deafferentation, the phenomenon is termed *orthograde transneuronal degeneration* (Fig. 8.2).

Retrograde Transneuronal Degeneration

Neurons can lose their normal targets without having their axons directly damaged in several ways. For example, if a lesion leads to retrograde degeneration of one set of neurons, the inputs to these neurons would be deprived of targets. Similarly, as noted above, if disease processes affect certain populations of neurons, the inputs to these cells would lose their normal targets.

The principles that determine the extent of retrograde transneuronal degeneration are similar to those that affect retrograde degeneration (see above). Although direct damage to the axon of the cells undergoing retrograde transneuronal degeneration is *not* a factor (by definition), *the extent of target loss is important; neurons with collateral projections are relatively resistant to retrograde transneuronal degeneration, whereas neurons projecting primarily to the target that is lost are most sensitive.* Thus, for retrograde transneuronal degeneration the concept of sustaining collaterals applies in the same way as for retrograde degeneration.

◁————————————————————————————————

FIGURE 8.2. Secondary degeneration of neurons after injury. For each panel, the normal circuitry is illustrated on the left; the initial damage produced by the lesion is illustrated in the middle panel, and the secondary degeneration that results from the lesion is illustrated on the right. *A* When the sole axonal projection of a neuron is damaged, its cell body may degenerate (retrograde degeneration). Also, neurons that lose most of their inputs often undergo orthograde transneuronal degeneration. Neurons that lose their normal targets may also degenerate (retrograde transneuronal degeneration). Orthograde and retrograde transneuronal degeneration occur even though the neurons undergoing the secondary degeneration are not directly damaged by the lesion. *B* When one axon branch is damaged, but a second branch is spared, the neuron is less likely to undergo retrograde degeneration, but it usually exhibits retrograde atrophy. The surviving collateral sustains the neuron, and is thus termed a *sustaining collateral*. *C* When a neuron loses input from one source, but retains input from some other source, it is less likely to undergo orthograde transneuronal degeneration. In this case, the surviving input can be considered a *sustaining alternate input*, operating in a manner that is similar to the sustaining collaterals in *B*. Partially denervated neurons may nevertheless exhibit orthograde transneuronal atrophy.

Retrograde degeneration due to target loss is thought to occur as a consequence of the loss of trophic support from the target. Nerve growth factor-dependent systems are the best understood. Such cells can be "saved" from retrograde degeneration by delivering NGF artificially (see Chapter 5). There is growing evidence for the existence of retrograde factors for other types of neurons, and it may well eventually be possible to prevent retrograde degeneration by treatment with such factors.

Orthograde Transneuronal Degeneration

Some neurons also degenerate when deprived of their normal synaptic input (see Fig. 1.3). In general, the severity of the transneuronal degeneration depends on the extent of the denervation and on developmental age. In fact, most systems exhibit orthograde transneuronal degeneration only during early development.

Again, neurons that retain some synaptic inputs (from undamaged sites) are more resistant to orthograde transneuronal degeneration than those whose inputs are completely eliminated. In fact, unless the removal of input is virtually complete, most cells exhibit *transneuronal atrophy* (see below) rather than transneuronal degeneration. This provides a corollary to the concept of sustaining collaterals: *the elimination of synaptic input is likely to result in orthograde degeneration of the postsynaptic cell if that cell receives few other inputs*. This is the *concept of sustaining alternate inputs*.

Transneuronal Atrophy

For most neurons the effects of denervation or target loss are not severe enough to lead to the death of the cell. Such cells may nevertheless exhibit regressive phenomena.

Retrograde Transneuronal Atrophy

Cells that do not die after axotomy or target removal often exhibit a biphasic response. In the early period after the injury, there is a dispersal of Nissl substance, the nucleus often becomes eccentric, and the neuronal soma swells. This response is termed *chromatolysis* or *the retrograde cell response*. During this early phase the neuron may exhibit some axonal regeneration, and some of the changes associated with chromatolysis may, in fact, reflect growth processes rather than regressive events.

An interesting phenomenon that is often observed during the chromatolytic phase is *bouton shedding* (Fig. 8.3). In this phenomenon the synaptic inputs to the axotomized cell disconnect. The synapses remain in the region of the cell body, but are separated from it by glial processes. The result of this disconnection is a profound depression of synaptic transmission to the affected cell. As a result, collateral projections that are not directly affected by the damage may be functionally disrupted.

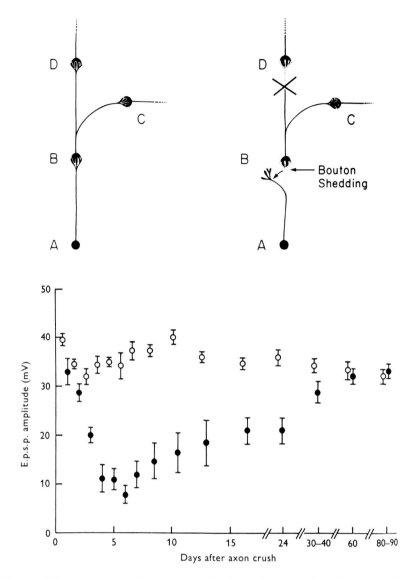

FIGURE 8.3. Amputation of an axon can lead to a disconnection of synapses from the cell body of the axotomized cell (upper figure). This is termed "bouton shedding." After the cut axon regenerates, the disconnected synapses usually reconnect. The lower graph illustrates the time course of this process in the superior cervical ganglion after a crush injury of postganglionic nerves. Synaptic efficacy is measured by recording intracellularly after stimulation of the preganglionic axons. *Filled circles* reveal the average excitatory postsynaptic potential (EPSP) amplitude in ganglion cells at various times after the crush. *Open circles* illustrate the average EPSP amplitude from the control ganglia contralateral to the crush. [The lower graph is from Purves D (1975) Functional and structural changes in mammalian sympathetic neurons following interruption of their axons. *J Physiol* 252: 429–463.]

If there is successful regeneration of the transected axon, the changes in the cell body are reversed, and the cell returns to near-normal morphology. If regeneration is not successful (in terms of regeneration to the normal target site), the cell may still survive. To some extent, this survival may be mediated through the establishment of alternative projections (see *regenerative sprouting* below). The neuron may also survive without establishing new connections if it has sustaining collaterals.

Neurons that survive axotomy may be somewhat atrophic. This is reflected by decreases in the size of the cell body; in metabolism, including protein synthesis; and in the production of neurotransmitter enzymes. These changes may have important consequences for transmission via surviving collaterals.

Orthograde Transneuronal Atrophy

Cells that survive deafferentation may also exhibit atrophic responses. *In general, the extent of the atrophy is related to the extent of the deafferentation*. Thus, cells receiving innervation from predominantly one source are most severely affected. Those that survive may exist in a severely atrophic state, reflected again by decreases in cell size and cell metabolism and by changes in second-order projections (emanating from the atrophied cells). Examples of severe atrophy following denervation include muscles and neurons in sensory relay nuclei in the CNS. For example, after removal of one eye, neurons in the lateral geniculate nucleus (LGN) undergo severe atrophy (Fig. 8.4). If the eye is removed during early development, the projections of these LGN neurons to the cerebral cortex are dramatically reduced. Similar transneuronal atrophy occurs in other sensory relay nuclei after disruption of input from the periphery.

In general, if the deafferentation is restricted to a portion of the post-synaptic cell's receptive surface, the atrophic changes will be restricted to the portion of the receptive surface that is denervated. Thus, deafferentation may lead to the loss of spines, the loss of dendritic branches, and even the loss of entire dendrites (Figs. 8.5 and 8.6). Again, the concept of sustaining alternate inputs applies. Thus, if individual dendrites are partially denervated, they usually survive; if denervation is complete, the entire dendrite may disappear.

Regressive effects due to injury are, in general, more severe in young animals. Some neurons that die when deprived of their inputs early in development survive without apparent ill effects when the same deafferentation occurs at maturity. In the auditory system of the chick, for example, there is a rapid loss of neurons in the avian homologue of the cochlear nucleus if the eighth nerve is damaged during the first few weeks of life, but not if a similar injury occurs in mature birds. Similarly, retrograde degeneration is usually more severe in young animals. The reasons for the greater sensitivity of developing tissue are poorly understood, but the fact has important implications for understanding the sequellae of

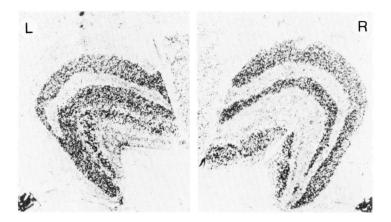

FIGURE 8.4. Orthograde transneuronal atrophy of neurons in the lateral geniculate nucleus (LGN) following removal of the right eye of a 2-week-old monkey. *L* LGN on the left side of the brain; *R* LGN on the right. Projections from each eye terminate in alternating laminae in the LGN. Removal of one eye leads to an atrophy of the neurons in the denervated laminae. [From Hubel DH, Wiesel TN, LeVay S (1977) Plasticity of ocular dominance columns in the monkey striate cortex. *Philos Trans R Soc Lond Biol* 278: 377–409.]

damage in the young. Indeed, injuries during early development may lead to *cascading degeneration* at sites that are physically quite separate from the actual site of injury.

Adaptive Changes in Response to Trauma

Denervation Supersensitivity

One of the frequent consequences of denervation is an increase in sensitivity to the transmitter utilized by the input that has been eliminated. This phenomenon is termed *denervation supersensitivity*. The best understood example of denervation supersensitivity is at the neuromuscular junction. In the normal muscle, sensitivity to ACh is restricted to the portion of the postsynaptic membrane just under the synaptic contact. After denervation, the entire muscle becomes sensitive. This comes about as a result of the appearance of new ACh receptors throughout the muscle fiber (Figs. 8.7 and 8.8).

The development of supersensitivity seems to be a consequence of changes in impulse traffic rather than a disruption of the delivery of some trophic substance from nerve to muscle. This is indicated by several lines of evidence. First, new ACh receptors appear after blocking of nerve conduction, even though the nerve is still in place (Fig. 8.7). Second, the increases in receptors can be prevented by direct electrical stimulation of the muscle.

Supersensitivity as the result of the development of new receptors is not unique to muscle. Identical responses are observed in ganglion cells of the peripheral nervous system, and there is evidence to suggest similar responses in the CNS. The development of supersensitivity can be considered an adaptive response since it may contribute to the reestablishment of transmission if the denervation is partial. These effects may be partic-

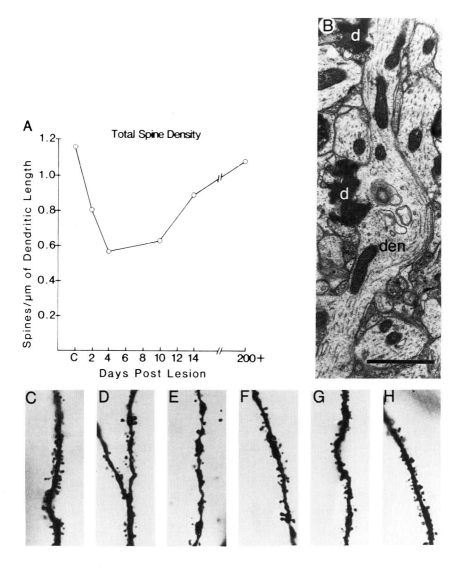

ularly important in the CNS, where most neurons receive inputs from multiple sources.

The appearance of new receptors may make the postsynaptic cell receptive to new innervation. For example, when muscles are chronically paralyzed with botulinum toxin, new receptors appear throughout the muscle fiber. In response to some signal presumably derived from the muscle, the nerve innervating the paralyzed muscle often sprouts so that the extent of innervation of the paralyzed muscle increases. If the disruption of activity is partial, the increases in sensitivity of the postsynaptic cell coupled with the compensatory growth of presynaptic process can serve to increase the degree of transmission, partially compensating for the deficits.

Reactive Growth of Axons and Synapses

Axonal regeneration involves the regrowth of a damaged axon to its original target. It is useful to distinguish between *regeneration* and *regenerative sprouting* (see below). In complete regeneration, the axon successfully reconnects with its normal target. With regenerative sprouting, there is a growth of the damaged axon, but the growth does not lead to reconnection with the normal target. Although axonal regeneration occurs to some extent in the peripheral nervous system, it is quite rare in the CNS, except in lower vertebrates such as fish and frogs.

The absence of true axonal regeneration in the CNS is probably not due to an inability of CNS neurons to grow. Indeed, CNS neurons are capable of responding vigorously to denervation by elaborating new syn-

---

◁—————————————————————————————

FIGURE 8.5. Loss of spines with denervation and their reappearance with reinnervation. The dendrites of granule cells of the dentate gyrus were evaluated after destruction of the input from the entorhinal cortex. Lesions of the entorhinal cortex remove about 90% of the synapses terminating on distal dendrites of granule cells (see Fig. 8.11). Over time, these dendrites are reinnervated as a result of sprouting. *A* Counts of the number of spines per unit length of dendrites reveal a loss of spines as existing inputs degenerate, followed by a reappearance of spines with reinnervation. *B* Electron microscopic studies reveal that the spines appear to collapse into the dendrites, resulting in a beaded appearance of the dendrites. *C–H* Photomicrographs of granule cell dendrites in normal animals (*C*) and at 2, 8, 14, 30, and 60 days postlesion respectively. [*A* is from Steward O (1986) Lesion-induced synapse growth in the hippocampus, in search of cellular and molecular mechanisms, in Isaacson RL, Pribram KH (eds): *The Hippocampus*, vol 3. New York, Plenum Press, pp 65–111; *B–H* is from Caceres AO, Steward O (1983) Dendritic reorganization in the denervated dentate gyrus of the rat following entorhinal cortical lesions. A Golgi and electron microscopic analysis. *J Comp Neurol* 214: 387–403.]

A

B

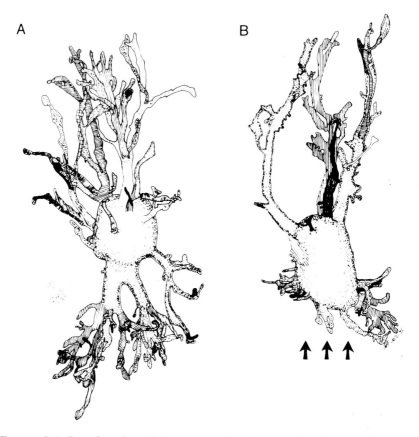

FIGURE 8.6. Deterioration of denervated dendrites. *A* Drawing of a Golgi impregnated neuron from nucleus laminaris of the chicken (the avian homologue of the medial superior olive of mammals). These neurons normally give rise to symmetrical dendritic arbors that emerge from dorsal and ventral poles of the cell. *B* illustrates a neuron from a similar location in the nucleus 10 days after removal of the normal innervation of the ventral set of dendrites. Note the disappearance of the dendrites on the ventral side (*arrows*). (Courtesy of J.S. Dietsch, Z.D.J. Smith, and E.W. Rubel.)

apses (see *sprouting* below). Furthermore, axons from CNS neurons can grow for long distances through peripheral nerve grafts (see below). Thus, other factors are thought to limit successful axonal regeneration. Some of the ones that are consistently mentioned, particularly with regard to the absence of regeneration following spinal cord injury, are the following:

1. The formation of a glial scar. Glia proliferate in the mature CNS in response to injury and form a *glial scar*. In the spinal cord such scars are formed of both glial elements and connective tissue. It has been

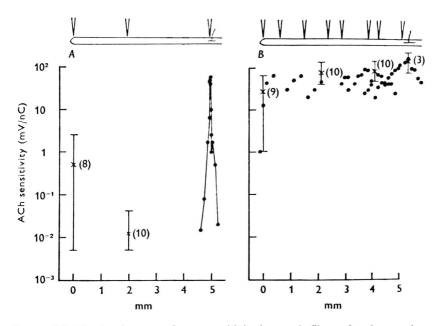

FIGURE 8.7. The development of supersensitivity in muscle fibers after denervation. ACh sensitivity along normally innervated muscles is restricted to the end-plate region. This can be documented by mapping ACh sensitivity along the muscle by applying ACh iontophoretically at different locations. The upper portion of the figure illustrates the experimental arrangement. The position of the end-plate is on the right side of the muscle fiber. Note the localized sensitivity at the end-plate region in the normal muscle. When the nerve is cut, or activity over the nerve is silenced, the entire muscle becomes sensitive to iontophoretically applied ACh. The graph on the right illustrates the high sensitivity along the entire length of the muscle when nerve activity had been blocked for seven days by a cuff containing local anesthetic. Similar results are obtained following denervation. [From Lømø T, Rosenthal J (1972) Control of ACh sensitivity by muscle activity in the rat. *J Physiol* 221: 493–513.]

suggested that such scars pose an actual physical barrier to regenerating fibers. There is little doubt that glial scars, particularly the glial/connective tissue scars in the spinal cord, are inhospitable territory for growing axons. Nevertheless, glial scars do not explain the absence of axonal regeneration in the CNS, since axon regeneration is not improved when scar formation is prevented.

2. Exposure to blood-borne toxins or antibodies. The CNS is buffered from many chemicals in the blood by the blood-brain barrier. In addition, the CNS is normally an immunological preserve, since antibodies also respect the barrier. Trauma disrupts the barrier and exposes the nervous system to blood-borne agents at the site of the injury. It has been suggested that such blood-borne products interfere with regen-

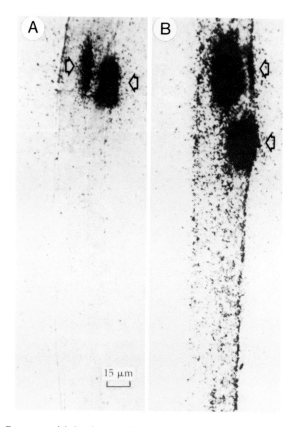

FIGURE 8.8. Supersensitivity in muscle occurs as a result of an increase in "extrajunctional" ACh receptors. *A* The distribution of ACh receptors in normally innervated muscle as revealed by autoradiography following binding of radioactively labeled α-bungarotoxin. Two muscle fibers from the diaphragm of an adult rat are illustrated, each with an end-plate region (*arrows*). Note that receptors are restricted to the end-plate region. *B* Distribution of ACh receptors in muscle fibers following denervation. Note the marked increase in silver grains in areas away from the junctional region. The end-plate region remains the area of highest receptor concentration. [From Fambrough DM (1981) Denervation: Cholinergic receptors of skeletal muscle, in Lefkowitz RJ (ed): *Receptors and Recognition Series*. London, Chapman and Hall, vol 13, pp 125–142.]

eration. Although breakdown of the blood-brain barrier at the site of injury may certainly contribute to an inhospitable environment, it again cannot completely account for the absence of regeneration. Grafts of peripheral nerve (see below) induce the growth of axons from the CNS into the graft, even though the blood-brain barrier is disrupted at the site of implantation.

3. Lack of supporting matrix or a favorable cellular environment. Axons of the peripheral nervous system (PNS) are capable of some regeneration, although it is not always effective because of errors. One difference between CNS and PNS axons is the supporting cells. Schwann cells ensheath axons of the PNS whereas oligodendrocytes ensheath CNS axons. Studies using nerve grafts indicate that PNS nerves fail to regenerate into grafts of optic nerve which contain only oligodendrocytes. This and other evidence suggest that Schwann cells provide substances or create an environment that is conducive to regeneration.
4. Ectopic synapse formation. Damaged axons of the CNS sometimes begin to grow, but form synaptic connections with inappropriate targets near the site of transection. By forming ectopic connections, it is thought that the stimulus for growth is removed, despite the fact that the cells have not reconnected with their normal targets. Similarly, reinnervation by inappropriate axons may make denervated postsynaptic cells unreceptive even if their normal inputs successfully regenerate.

Factors that seem to be related to the presence or absence of regeneration include developmental age and the nature of the anatomical system involved. In general, younger animals exhibit more vigorous reactive growth responses than mature animals. This is something of a paradox given the more extensive regressive events, but it is nevertheless true that, for whatever reason, recovery is usually much more complete if damage occurs early in life.

The extent of regeneration also depends on the anatomical system involved in that some systems may be more capable of regeneration than others. For example, CNS pathways using serotonin, norepinephrine, and ACh seem to exhibit greater regeneration than other systems. This may be because these systems do not require the same sort of precise connectivity as other pathways (see Chapter 3).

Regenerative Sprouting

Regenerative sprouting involves the growth of an axon that has been interrupted when the growth does *not* lead to reconnection with its original target. Regenerative sprouting may result in synapse formation on inappropriate elements, or may culminate in blindly ending axonal sprouts. The latter are thought to degenerate over time if they do not reconnect.

Regenerative sprouting is potentially important for several reasons. First, the existence of regenerative sprouting indicates that the failure of axon regeneration is not a result of an intrinsic inability of damaged neurons to elaborate new axons. Second, regenerative sprouting could contribute to functional recovery by establishing alternative pathways for the transmission of information. It is not known whether this sort of alternative pathway formation actually contributes to recovery. On the other hand, if ectopic connections are formed, regenerative sprouting could actually impede regeneration by eliminating the stimulus for growth (see above).

The stimulus for regenerative sprouting is not clear. In part, it may be related to the pruning of the axon. Neurons may seek to establish a certain total axon arbor, and when part of this arbor is lost, the neuron may compensate by growing additional connections. In this, regenerative sprouting may be closely related to the "pruning-related growth" discussed below.

Regenerative sprouting may also be related to the presence of nearby denervated targets. If regenerative sprouting in a particular setting is due to a signal from a nearby denervated target, then the growth may have the same cellular mechanism as collateral or paraterminal sprouting. Regenerative sprouting is somewhat more common than axonal regeneration. It certainly occurs in the ACh, serotonin, norepinephrine and dopaminergic systems in the mammalian brain. There is also indirect but convincing evidence for its occurrence in the spinal cord.

Collateral or Paraterminal Sprouting

Collateral or paraterminal sprouting involves the growth of axonal and/ or synaptic elements in response to injury that does not directly affect the cell participating in the growth response. Collateral sprouting involves the growth of additional presynaptic arbors from the preterminal axon. Paraterminal sprouting involves the growth from existing terminals (Fig. 8.9). In most settings, collateral and paraterminal sproutings probably occur simultaneously; this is particularly true in CNS pathways where the normal connections are *en passant*.

The distinction between regenerative sprouting versus collateral or paraterminal sprouting is important because in regenerative sprouting, there is direct damage to the neuron that is growing, whereas in collateral and paraterminal sprouting, the growing neuron is responding to some extrinsic cues (probably originating from denervated cells). Indeed, the final extent of collateral and paraterminal sprouting is closely related to the availability of synaptic sites as a result of denervation.

Sprouting in the CNS

In many cases CNS neurons that lose their normal synaptic inputs are reinnervated as a result of the short-range growth of nearby undamaged fiber systems. Most examples of sprouting in the CNS involve the formation of additional synaptic connections by undamaged axonal systems that are present in the denervated zone (Fig. 8.10). Because most examples of reinnervation involve the formation of new synapses without long-distance growth of axons, the reinnervation process is often termed *reactive synaptogenesis*.

Central nervous system sprouting has been observed in many systems, and there is good reason to believe that it is a common response to de-

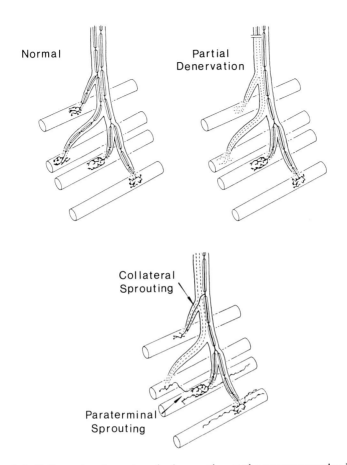

FIGURE 8.9. Collateral and paraterminal sprouting at the neuromuscular junction following partial denervation. When muscle fibers are denervated by cutting part of the nerve supply to a muscle group, the surviving axons respond by growing new connections to reinnervate the denervated muscle fibers. This growth can occur as a result of the sprouting of a collateral branch somewhere along an existing axon, usually at a node of Ranvier (collateral sprouting), or can occur as a result of the growth of a new terminal arbor from an existing presynaptic terminal (paraterminal sprouting). [Reproduced, with permission, from the Annual Review of Neuroscience, Vol. 4. © 1981 by Annual Reviews Inc.]

nervation. The time course of reinnervation varies to some extent across anatomical systems and species. One of the best studied systems is the hippocampal formation. Quantitative electron microscopic studies have revealed that in rat hippocampus, reinnervation begins between four and six days postlesion, and the growth continues at a rapid pace between about six and twelve days postlesion. Physiological studies reveal that pathways participating in the sprouting response increase their physiological potency over a comparable time course; presumably the increase

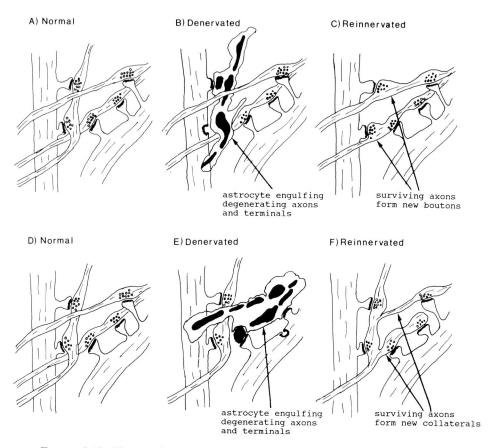

FIGURE 8.10. The development of collateral reinnervation in the CNS (reactive synaptogenesis). When CNS neurons are partially denervated, the synapses that degenerate are often replaced as a result of the elaboration of new synapses by surviving axons near the denervated sites. It is thought that such reactive synaptogenesis occurs as a result of local growth. Reinnervation may occur as a result of the formation of new presynaptic specializations along existing axons (A–C) or as a result of relatively short-distance collateral sprouting similar to that which occurs at the neuromuscular junction. (D–F). [A–C after McWilliams R, Lynch G (1978) Terminal proliferation and synaptogenesis following partial deafferentation: the reinnervation of the inner molecular layer of the dentate gyrus following removal of its commissural afferents. *J Comp Neurol* 180: 581–616.]

in synaptic potency reflects the establishment of new synapses by the sprouting pathways (Fig. 8.11). Identical growth responses occur in other species, including cats and monkeys, and there is some evidence for post-lesion growth in humans. Interestingly, the time course of growth seems to be somewhat slower in cats and monkeys.

There is evidence that sprouting may contribute to recovery of function following brain injury. For example, damage to hippocampal pathways

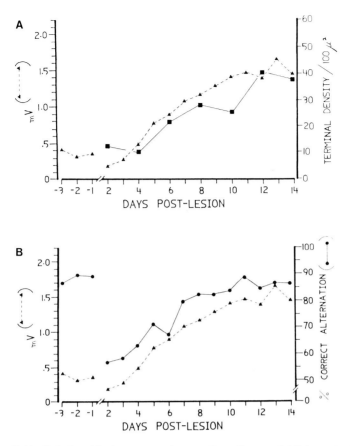

FIGURE 8.11. Relationship between reinnervation, the reestablishment of physiologically operational synapses, and recovery of function following CNS injury. *A* The time course of terminal proliferation in the dentate gyrus of the rat following unilateral lesions of the entorhinal cortex is compared with the time course of the establishment of operational synapses by one of the pathways that participates in the sprouting response. The time course of reinnervation can be defined by counting the number of presynaptic terminals and synapses in denervated areas as a function of time after the lesion. The graph illustrates such counts in the molecular layer of the dentate gyrus at various times after destruction of the normal input to the molecular layer from the ipsilateral entorhinal cortex. Physiological operation of one of the pathways that participates in the sprouting response was assessed by measuring the maximal-evoked response that could be generated by the pathway at various times after the lesion. *B* The time course of reestablishment of operational synapses is compared with the time course of recovery of function as evaluated by performance in a T-maze task that measures short-term memory for spatial cues. Note that recovery of function occurs with roughly the same time course as the reinnervation. [From Reeves T (1983) Electrophysiological correlates of recovery from entorhinal lesions. Ph.D. thesis, Department of Psychology, Southern Illinois University. The plots of terminal density are from Steward O, Vinsant SV (1983) The process of reinnervation in the dentate gyrus of the adult rat: A quantitative electron microscopic study of terminal proliferation and reactive synaptogenesis. *J Comp Neurol* 214: 370–386.]

leads to well-characterized deficits in certain tasks that require memory based on spatial cues. These deficits recover after injury, and the time course of recovery closely parallels the time course of reinnervation of neurons in the hippocampus (Fig. 8.11).

Another well-characterized sample of sprouting occurs in the dorsal root projections to the spinal cord after injury to the cord itself or after injury to other dorsal roots. The sprouting of primary afferents is thought to contribute to the development of hyperreflexia. To some extent, hyperreflexia is maladaptive, but it can also contribute to recovery of limb movement. In corticospinal pathways, the establishment of abnormal ipsilateral pathways following cortical damage in young animals may permit some degree of use of limbs that would normally be controlled by the damaged side of the brain. There is some evidence that a similar reorganization of circuitry contributes to spared function after cortical damage in developing humans. For example, children who suffer unilateral injuries to the portion of the motor cortex supplying the limbs often show sparing of function at maturity. However, motor activities in limbs that would normally be controlled by the damaged side are sometimes accompanied by *mirror movements* on the opposite side. These mirror movements may reflect the fact that cortical pathways that are normally unilateral have bilateral components as a result of postlesion reorganization. The fact that lesions often lead to less impairment if they occur early in development may be related to the increased capacity of the young brain to reorganize its pathways.

Pruning-related Sprouting

The final form of naturally occurring growth following injury is termed *pruning-related sprouting*. Pruning-related sprouting refers to the growth of axonal and/or synaptic connections when one collateral of the sprouting cell is damaged. The distinction between pruning-related sprouting and regenerative sprouting is that pruning-related sprouting need not involve the axon that suffers the damage. Thus, growth can occur at sites that are remote from the injury (see Fig. 8.1).

Pruning-related sprouting may be particularly significant when there is damage to neurons with wide-ranging collaterals. Thus, damage to the rostrally directed projections of cells in the locus ceruleus leads to compensatory sprouting of the collaterals of the damaged cells in the cerebellum.

Principles

The incidence, extent, and speed of sprouting depend on a number of factors:

1. Sprouting of all types is more likely to occur in young animals. Some examples of sprouting are found exclusively after lesions early in development. It is important to recall, however, that some of these may

represent projections that are actually present at the time of the injury; thus, the reorganization that results from injury may actually represent a retention of a projection that is normally present early in development rather than the growth of an abnormal pathway (see Chapter 7).

2. Sprouting occurs much sooner after injury in young animals. This may be because neurons in young animals are still actively growing, whereas neurons in mature animals must be induced to grow.

3. In mature animals, sprouting is a close-range phenomenon that depends on *proximity* of the surviving fibers to the denervated site. Thus, sprouting may be particularly important when there is partial damage to a pathway, or when neurons receive inputs from multiple sources that have overlapping terminal fields.

4. To some extent, sprouting depends on the relationship between the fiber that is available to provide the reinnervation and the normal system that has been damaged. Reinnervation is much more likely if the systems are of similar type (homologous); presumably, this reflects the ability of the postsynaptic cell to accept innervation from different systems.

The extent of reinnervation may also depend on the temporal relationship between transneuronal degeneration and the beginning of any sprouting response. Denervation usually leads to some transneuronal atrophy, but in some systems, the denervated dendrites are preserved long enough to be reinnervated. In other systems there is extensive transneuronal atrophy soon after the disruption of the afferents to the cells. Such rapid transneuronal atrophy may prevent reinnervation from occurring. For example, in the chick, the removal of inputs to one set of dendrites of the neurons in the medial superior olive leads to the degeneration of those dendrites within a few days (Fig. 8.6). Under these circumstances, there is no evidence of sprouting. However, when complete denervation is produced in stages, the dendrites are partially preserved, and in this case there is some sprouting. These results suggest that if cells can be prevented from undergoing transneuronal degeneration, reinnervation may eventually occur. It is reasonable to anticipate future intervention strategies that capitalize on this notion.

Transplants and Grafts

Over the last few years dramatic progress has been made in the use of transplants. There have been two principal strategies. The first strategy involves the use of peripheral nerve grafts to bridge across areas of damage in order to facilitate regeneration of axons in the host CNS. The second strategy involves the use of transplants as actual replacement tissues.

Transplants and Grafts as Aids to Regeneration

The rationale of the approach is that the CNS environment is not conducive to axonal regeneration, whereas some feature of the PNS environment is. Thus, peripheral nerve grafts have been used as bridges in an attempt to stimulate the growth of CNS axons. The long-range goal is to find a means to promote reconnection, particularly in the spinal cord. Experiments by a number of investigators, particularly Albert Aguayo and his colleagues, reveal that CNS neurons can regenerate for long distances through such grafts (Aguayo et al. 1982a,b).

Aguayo et al. (1981) inserted peripheral nerve grafts into the medulla and the thoracic spinal cord of rats, and allowed the grafts to remain in place for several weeks (Fig. 8.12). They evaluated whether any axons had grown into the graft by cutting the graft, and soaking each cut end in a solution of horseradish peroxidase which is taken up and transported by cut axons. In this way, they labeled the axons that entered the graft, and the cells of origin of the axons. They found that the graft stimulated a dramatic ingrowth of axons from neurons in the medulla and spinal cord; in fact, the axons appeared to extend for the full length of the graft and then re-entered the host CNS (Fig. 8.12).

More recent studies have used a similar strategy to actually reconnect the eye with the normal visual centers in the brain after damage to the optic nerve in rats. For these studies, the optic nerve was transected, and replaced by a segment of autologous (from the same animal) peripheral nerve. One end of this peripheral nerve graft was apposed to the orbital stump of the optic nerve (near where the optic nerve exited from the eyeball). The opposite end of the nerve graft was ligated and temporarily placed between the skin and the scalp. One to two months after making the initial anastomosis, the opposite end of the graft was inserted into the CNS near the border between the lateral geniculate nucleus and the superior colliculus (two normal targets of optic nerve axons). The investigators found that axons from the eye successfully regenerated through the peripheral nerve graft, and even succeeded in forming synaptic connections with neurons in the superior colliculus (Fig. 8.13). This extraordinary study clearly demonstrates that peripheral nerve grafts can promote the regeneration of CNS axons, and suggests that grafts may eventually provide the means to promote reconnection of neurons with their normal targets.

Transplants as Tissue Replacements

A second promising approach involves the replacement of damaged tissue with similar tissue obtained from fetal animals. The CNS is a favorable site for transplants, and brain tissue obtained from developing animals survives and differentiates within the host CNS. Most importantly, neurons within the transplants establish connections with neurons of the host CNS.

Transplants have been used with considerable success in experimental animals to reconstruct neural pathways that have been destroyed by lesions. Studies have revealed that the connections formed between transplant and host are electrophysiologically functional, and animals with transplants exhibit a greater postinjury recovery than those without.

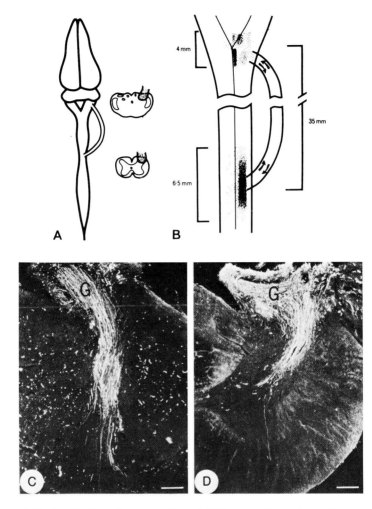

FIGURE 8.12. Facilitation of regeneration of CNS axons by peripheral nerve grafts. A peripheral nerve graft was inserted as a bridge between the medulla and the thoracic spinal cord of the rat. After sufficient time was allowed to permit axonal growth, HRP was applied to the graft to retrogradely label neurons that send axons into the graft. The *dots* in *B* indicate the position of retrogradely labeled neurons in the medulla and spinal cord. [From David S, Aguayo AJ (1981) Axonal elongation into peripheral nervous system "bridges" after central nervous system injury in adult rats. *Science* 214: 931–933. Copyright 1981 by the AAAS.]

The work that has been important in terms of its potential clinical application has involved transplants of dopamine-containing neurons from the substantia nigra after destruction of the nigrostriatal pathways of the host. Parkinson's disease is characterized by a loss of nigro-striatal neurons; thus, a therapy that could replace these neurons is of considerable

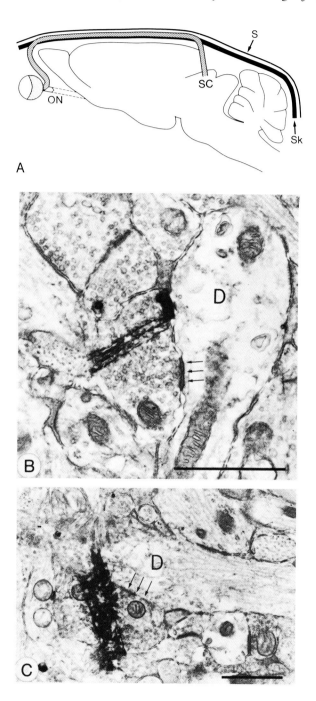

interest. In rats unilateral destruction of the substantia nigra leads to motor deficits that can be easily revealed by challenging animals with doses of amphetamine. Because of the asymmetry of the pathways, such treatments induce rotational behavior that normally does not recover (Fig. 8.14). When animals receive transplants of fetal substantia nigra into the denervated caudate nucleus, or injections of dissociated cells from fetal substantia nigra, the transplants elaborate extensive connections with the host caudate nucleus (Fig. 8.14). In animals bearing transplants, the amphetamine-induced rotational behavior recovers.

Similar striking recovery is observed in primates with experimentally induced Parkinson's disease. In these studies, monkeys were treated with the neurotoxin 1-methyl-4-phenyl-1,2,3,6-tetrahydropyridine (MPTP), a drug that selectively destroys neurons of the substantia nigra. This drug, initially manufactured illicitly as a "designer drug," was found to induce severe Parkinsonian symptoms in some users. When animals with MPTP-induced Parkinson's symptoms received injections of suspensions of substantia nigra neurons into the caudate nucleus, there was a progressive amelioration of the symptoms as the transplants established connections with the host CNS.

There is great enthusiasm that transplants might provide the means to replace tissue lost to injury and disease; already attempts are being made to apply transplant methodology in humans. In one approach, patients with Parkinson's disease have received autologous transplants of adrenal tissue. The rationale for this approach is that adrenal glands produce dopamine in addition to adrenalin. The first surgeries carried out in Sweden were disappointing because there was little evidence of functional improvement when the adrenal tissue was placed directly into the caudate nucleus. Subsequently, a slightly different strategy was attempted in a series of patients in Mexico City; in these patients, the adrenal tissue was placed on the parenchyma overlying the caudate nucleus. The reports from these patients were initially encouraging, prompting similar surgeries in the United States and elsewhere. Unfortunately, the results have been much less encouraging than the initial studies suggested.

◁

FIGURE 8.13. Facilitation of optic nerve regeneration by peripheral nerve grafts. A Peripheral nerve grafts were used to replace a transected optic nerve (ON) in a mature rat. One end of the autologous graft was attached to the orbital stump of the ON close to the eye. The other end of the graft was ligated, and left temporarily between the scalp (S) and skull (Sk). Two to three months later, the distal end of the graft was inserted into the superior colliculus (SC). B, C Electron micrographs of HRP-labeled presynaptic terminals in the superior colliculus labeled after injection of HRP into the peripheral nerve grafted eyes. Arrows indicate the synaptic contact site. D = dendrite; calibration bar = 1 μm. [From Vidal-Sanz M, Bray GM, Villegas-Perez MP et al. (1987) J Neurosci 7: 2894–2909. Copyright, Society for Neuroscience.]

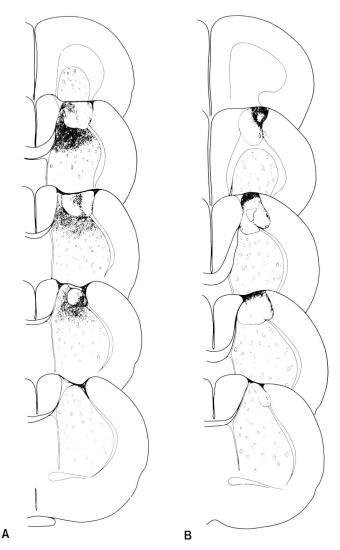

A   B

FIGURE 8.14. Growth of axons from fetal transplants of substantia nigra into the caudate nucleus of a host rat. The drawings illustrate two substantia nigra (SN) transplants 7 months after transplantation. The ingrowth of dopamine fibers (as revealed by histofluorescence) is represented by the *fine lines* and *dots*. There was extensive ingrowth of axons from the transplant to the caudate nucleus in the example illustrated in *A*; in this animal, there was also substantial behavioral recovery (see Fig. 8.15). The transplant illustrated in *B* also gave rise to axons that grew out of the transplant, but the ingrowth of axons from the transplant into the caudate nucleus of the host was minimal. In this animal there was no behavioral recovery. [From Bjorklund A, Dunnett SB, Stenevi U et al. (1980) Reinnervation of the denervated striatum by substantia nigra transplants: functional consequences as revealed by pharmacological and sensorimotor testing. *Brain Res* 199: 307–333.]

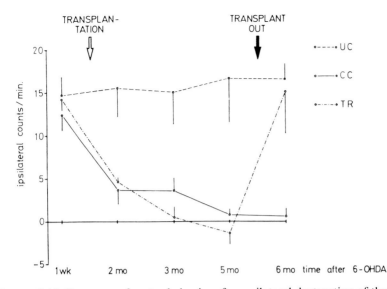

FIGURE 8.15. Recovery of motor behavior after unilateral destruction of the substantia nigra (SN) in animals receiving transplants of fetal SN. With unilateral SN lesions, rats exhibit uncontrolled rotational behavior when injected with amphetamine. There was no recovery in animals that did not receive transplants. In some of the animals that received transplants, the rotational behavior subsided over time (CC). In the animals exhibiting compensation, there was substantial ingrowth of dopamine fibers from the transplant into the caudate nucleus of the host (see Fig. 8.12A). Some of the animals receiving transplants remained uncompensated (UC); in these animals the extent of axonal ingrowth was minimal. In some of the "compensated" animals the transplant was removed after recovery had occurred. The uncontrolled rotational behavior in response to amphetamine challenge reappeared when the transplant was removed (TR). [From Bjorklund A, Dunnett SB, Stenevi U et al. (1980) Reinnervation of the denervated striatum by substantia nigra transplants: functional consequences as revealed by pharmacological and sensorimotor testing. *Brain Res* 199: 307–333.]

Early in 1988 the announcement was made by the group in Mexico City that patients with Parkinson's disease had received transplants of fetal tissue obtained from a spontaneously aborted fetus. The results of fetal transplants must still be evaluated, but judging from the work in experimental animals, there is every reason to be optimistic that the transplants may result in at least some functional recovery.

Research in transplants is moving at an astounding pace, and there is every reason to expect dramatic new advances. Perhaps the principal issues in the near-term will be more medico-legal than scientific. It is clear that the most effective transplants are those derived from fetal tissue. Unless it proves feasible to use alternative donors, there will be major issues to be resolved regarding the use of fetal human tissue for transplant therapy.

Supplemental Reading

Orthograde and Retrograde Degeneration

Cowan WM (1970) Anterograde and retrograde transneuronal degeneration in the central and peripheral nervous system, in Ebbesson SOE, Nauta WJH (eds): Contemporary Research Methods in Neuroanatomy. New York, Springer-Verlag, pp 217–251

Caceres A, Steward O (1983) Dendritic reorganization in the denervated dentate gyrus of the rat following entorhinal cortical lesions: a Golgi and electron microscopic analysis. *J Comp Neurol* 214:387–403

Powell TPS, Erulkar SD (1962) Transneuronal cell degeneration in the auditory relay nuclei of the cat. *J Anat* 91:249–268

Purves D (1975) Functional and structural changes in mammalian neurones following interruption of their axons. *J Physiol* 252:429–463

Tower SS (1939) The reaction of muscle to denervation. *Physiol Rev* 19:1–48

Denervation Supersensitivity

Cannon WB *Denervation Supersensitivity*. London, Macmillan.

Axelsson J, Thesleff S (1959) A study of supersensitivity in denervated mammalian skeletal muscle. *J Physiol* 147:178–193

Jones R, Vrbova G (1974) Two factors responsible for the development of denervation hypersensitivity. *J Physiol* 236:517–538

Kuffler SW, Dennis MJ, Harris, AJ (1971) The development of chemosensitivity in extrasynaptic areas of the neuronal surface after denervation of parasympathetic ganglion cells in the heart of the frog. *Proc R Soc Lond* 177:555–563

Lømø T, Rosenthal J (1972) Control of ACh sensitivity by muscle activity in the rat. *J Physiol* 221:493–513

Miledi R (1960) The acetylcholine sensitivity of frog muscle fibers after complete or partial denervation. *J Physiol* 151:1–23

Bouton Shedding

Matthews MR, Nelson VH (1975) Detachment of structurally intact nerve endings from chromatolytic neurons of rat superior ganglion during the depression of synaptic transmission induced by postganglionic axotomy. *J Physiol* 245:91–135

Role of Schwann Cells in Axon Regeneration

Aguayo A, David S, Richardson P et al. (1982a) Axonal elongation in peripheral and central nervous system transplants. *Adv Cell Neurobiol* 3:215–234

Aguayo AJ, Richardson PM, Benfey M (1982b) Transplantation of neurons and sheath cells-a tool for the study of regeneration, in Nicholls JG (ed): *Repair and Regeneration of the Nervous System,* Life Sciences Research Report 24. Berlin, Springer-Verlag, pp 91–106

Benfey M, Aguayo AJ (1982) Extensive elongation of axons from rat brain into peripheral grafts. *Nature* 296:150–152

Astrocytes and Phagocytosis of Degeneration Debris

Gall C, Rose G, Lynch G (1979) Proliferative and migratory activity of glial cells in the partially deafferented hippocampus. *J Comp Neurol* 183:539–550

Sprouting

Edds MV (1953) Collateral nerve regeneration. *Q Rev Biol* 28:260–276

Cotman CW, Nadler JV (1981) Reactive synaptogenesis in the hippocampus, in Cotman CW (ed): *Neuronal Plasticity*. New York, Raven Press, pp 227–271

Cotman CW, Nieto-Sampedro M, Harris EW (1981) Synapse replacement in the nervous system of adult vertebrates. *Physiol Rev* 61:684–784

Raisman G (1969) Neuronal plasticity in the septal nuclei of the adult rat. *Brain Res* 14:25–48

Steward O, Vinsant SV (1983) The process of reinnervation in the dentate gyrus of the adult rat: a quantitative electron microscopic analysis of terminal proliferation and reactive synaptogenesis. *J Comp Neurol* 214:370–386

Trophic Factors in CNS Regeneration and Repair

Nieto-Sampedro M, Manthorpe M, Barbin G et al. (1983) Injury-induced neuronotrophic activity in adult rat brain: correlation with survival of delayed implants in the wound cavity. *J Neurosci* 3:2219–2229

Skene JHP, Willard M (1981) Axonally transported proteins associated with axon growth in rabbit central and peripheral nervous systems. *J Cell Biol* 89:96–103

Skene JHP, Willard M (1981) Characteristics of growth-associated polypeptides in regenerating toad retinal ganglion cell axons. *J Neurosci* 1:419–426

Index

,